Responsible Drinking

ISSUES

Volume 194

Series Editor
Lisa Firth

Independence
Educational Publishers
Cambridge

First published by Independence
The Studio, High Green
Great Shelford
Cambridge CB22 5EG
England

Copyright

Photocopy licence

British Library Cataloguing in Publication Data

Responsible drinking. -- (Issues ; 194)

1. Alcoholism.

I. Series II. Firth, Lisa.

362.2'92-dc22

ISBN-13: 978 1 86168 555 1

Printed in Great Britain

MWL Print Group Ltd

CONTENTS

Chapter 1 Drinking Trends

Chapter 2 Problem Drinking

OTHER TITLES IN THE ISSUES SERIES

For more on these titles, visit: www.independence.co.uk

Food and Nutrition ISBN 978 1 86168 289 5
Energy Matters ISBN 978 1 86168 305 2
Exploited Children ISBN 978 1 86168 313 7
Stress and Anxiety ISBN 978 1 86168 314 4
Transport Trends ISBN 978 1 86168 352 6
Gambling Trends ISBN 978 1 86168 375 5
Customers and Consumerism ISBN 978 1 86168 386 1
A Genetically Modified Future? ISBN 978 1 86168 390 8
The Education Problem ISBN 978 1 86168 391 5
Vegetarian and Vegan Diets ISBN 978 1 86168 406 6
Mental Health and Wellbeing ISBN 978 1 86168 407 3
Media Issues ISBN 978 1 86168 408 0
The Cloning Debate ISBN 978 1 86168 410 3
Sustainability and Environment ISBN 978 1 86168 419 6
The Terrorism Problem ISBN 978 1 86168 420 2
Religious Beliefs ISBN 978 1 86168 421 9
A Classless Society? ISBN 978 1 86168 422 6
Migration and Population ISBN 978 1 86168 423 3
Climate Change ISBN 978 1 86168 424 0
Euthanasia and the Right to Die ISBN 978 1 86168 439 4
Sexual Orientation and Society ISBN 978 1 86168 440 0
The Gender Gap ISBN 978 1 86168 441 7
Domestic Abuse ISBN 978 1 86168 442 4
Travel and Tourism ISBN 978 1 86168 443 1
The Problem of Globalisation ISBN 978 1 86168 444 8
The Internet Revolution ISBN 978 1 86168 451 6
An Ageing Population ISBN 978 1 86168 452 3
Poverty and Exclusion ISBN 978 1 86168 453 0
Waste Issues ISBN 978 1 86168 454 7
Staying Fit ISBN 978 1 86168 455 4
Drugs in the UK ISBN 978 1 86168 456 1
The AIDS Crisis ISBN 978 1 86168 468 4
Bullying Issues ISBN 978 1 86168 469 1
Marriage and Cohabitation ISBN 978 1 86168 470 7
Our Human Rights ISBN 978 1 86168 471 4
Privacy and Surveillance ISBN 978 1 86168 472 1
The Animal Rights Debate ISBN 978 1 86168 473 8
Body Image and Self-Esteem ISBN 978 1 86168 484 4
Abortion – Rights and Ethics ISBN 978 1 86168 485 1
Racial and Ethnic Discrimination ISBN 978 1 86168 486 8
Sexual Health ISBN 978 1 86168 487 5
Selling Sex ISBN 978 1 86168 488 2
Citizenship and Participation ISBN 978 1 86168 489 9
Health Issues for Young People ISBN 978 1 86168 500 1
Crime in the UK ISBN 978 1 86168 501 8
Reproductive Ethics ISBN 978 1 86168 502 5
Tackling Child Abuse ISBN 978 1 86168 503 2
Money and Finances ISBN 978 1 86168 504 9
The Housing Issue ISBN 978 1 86168 505 6
Teenage Conceptions ISBN 978 1 86168 523 0
Work and Employment ISBN 978 1 86168 524 7
Understanding Eating Disorders ISBN 978 1 86168 525 4
Student Matters ISBN 978 1 86168 526 1
Cannabis Use ISBN 978 1 86168 527 8
Health and the State ISBN 978 1 86168 528 5
Tobacco and Health ISBN 978 1 86168 539 1
The Homeless Population ISBN 978 1 86168 540 7
Coping with Depression ISBN 978 1 86168 541 4
The Changing Family ISBN 978 1 86168 542 1
Bereavement and Grief ISBN 978 1 86168 543 8
Endangered Species ISBN 978 1 86168 544 5
Responsible Drinking ISBN 978 1 86168 555 1
Alternative Medicine ISBN 978 1 86168 560 5
Censorship Issues ISBN 978 1 86168 558 2
Living with Disability ISBN 978 1 86168 557 5
Sport and Society ISBN 978 1 86168 559 9
Self-harming and Suicide ISBN 978 1 86168 556 8

EXPLORING THE ISSUES

Photocopiable study guides to accompany the above publications. Each four-page A4 guide provides a variety of discussion points and other activities to suit a wide range of ability levels and interests.

A note on critical evaluation

Because the information reprinted here is from a number of different sources, readers should bear in mind the origin of the text and whether the source is likely to have a particular bias when presenting information (just as they would if undertaking their own research). It is hoped that, as you read about the many aspects of the issues explored in this book, you will critically evaluate the information presented. It is important that you decide whether you are being presented with facts or opinions. Does the writer give a biased or an unbiased report? If an opinion is being expressed, do you agree with the writer?

Responsible Drinking offers a useful starting point for those who need convenient access to information about the many issues involved. However, it is only a starting point. Following each article is a URL to the relevant organisation's website, which you may wish to visit for further information.

DRINKING TRENDS

Young people and alcohol

Information from Alcohol Concern.

Introduction

Drinking among young people, in particular excessive drinking, is a major concern for parents, practitioners and the wider community.

There is increasing evidence of the impact drinking is having on young people's long- and short-term health, as well as their chances of being in risky situations when drunk. There are also implications for crime and anti-social behaviour as well as for society as a whole.

This article looks at the existing evidence on young people's drinking, showing the prevalence of drinking and highlighting alcohol-related problems that are specific to young people.

Drinking patterns and trends

In recent years there have been significant changes in the way young people drink and how much they drink. Overall the proportion of young people who do not drink is increasing. However, among those who do drink, there seems to have been an increase in alcohol consumption:

- The proportion of 11- to 15-year-olds across England who have never drunk alcohol has risen from 39% in 2003 to 48% in 2008.
- In 2008, over half (52%) of 11- to 15-year-olds had drunk at least one alcoholic drink in their lifetime. This increases with age from 16% of 11-year-olds to 81% of 15-year-olds.
- There has been a decline in the number of 11- to 15-year-olds who have drunk in the last week. In 2008, 18% of 11-15 year olds had drunk alcohol at least once a week, down from 26% in 2001. This

Law relating to alcohol and young people (summary).

Under five years old	It is illegal to give an alcoholic drink to a child under five except under medical supervision in an emergency.	Children and Young Person's Act 1933
Under 16 years old	Children under 16 are allowed on licensed premises as long as they are supervised by an adult, but cannot have any alcoholic drinks. However, some premises may be subject to licensing conditions preventing children from entering, such as pubs which have experienced problems with under-age drinking.	Licensing Act 2003
Under 18 years old	It is illegal for anyone under 18 to buy alcohol in a pub, off-licence, supermarket or other outlet or for anyone to buy alcohol for someone under 18 to consume in a pub or public place. The only exception is where young people aged 16 or 17 can drink beer, wine or cider with a table meal if it is bough by an adult and they are accompanied by an adult.	Licensing Act 2003
Under 18 years old	Police have powers to confiscate alcohol from: a) under-18s in possession of alcohol, in a public place; b) someone (in a public place) who intends that alcohol in their possession should be consumed by a person under the age of 18 in a public place.	Confiscation of Alcohol (Young Persons) Act 1997

Source: Alcohol Concern

ALCOHOL CONCERN

is similar for boys and girls. The proportion that had drunk in the last week increases with age from 3% of 11-year-olds to 38% of 15-year-olds. White pupils are more likely to have drunk alcohol recently than Black or Asian pupils.

- Of those who drank in the last week, the average weekly consumption has more than doubled since 1990 from 5.3 units per week to 11.4 units per week in 2006. In 2008, 11- to 15-year-olds who drank in the last week consumed an average of 14.6 units (and a median intake of 8.5 units), equivalent to over seven pints of normal strength beer or nearly one and a half bottles of wine. (The method used to calculate alcohol consumption changed in 2007 and it is not possible to compare 2008 consumption directly with that method in 2006 and earlier.)
- In 2008, boys tended to drink more than girls. Boys who drank in the last week drank more units of alcohol (16.0 units) than girls who drank in the last week (13.1 units).

11- to 15-year-old pupils in the survey were more likely to have drunk alcohol (52%) than to have smoked (32%) or tried drugs (22%)

- Older pupils who drank in the last week drank more than younger pupils. On average 15.5 units for 15-year-olds, compared with 12.0 units for 11- to 13-year-olds in 2008.
- Compared with pupils in London schools, those in Yorkshire and the Humber had increased odds of having drunk alcohol in the last week.
- A comparative European study of drinking among 15- to 16-year-old European students showed 88% of British students had consumed alcohol during the past 12 months and more than half (57%) had been drunk during the same period. The estimated consumption on the latest drinking day (6.2 cl 100%) is well above ESPAD mean.
- 11- to 15-year-old pupils in the survey were more likely to have drunk alcohol (52%) than to have smoked (32%) or tried drugs (22%).
- Binge-drinking patterns were more likely among drinkers from more deprived areas.

Survey findings can vary however. For example, the 2008 TellUs3 survey showed significant differences to the above data in terms of the prevalence of young people who are drinking. This survey indicates that 25% of pupils aged 10-15 said they had never had an alcoholic drink (compared to 46% of those in the sample from *Drug use, smoking and drinking among young people in England in 2007*).

It is important to note therefore that surveys offer a view on people's behaviour which may not always accurately reflect their actual behaviour.

Drinking styles – transitions in drinking

Newburn and Shiner in their 2001 literature review identified several stages in young people's drinking behaviour which changes with increasing age:

- 12- to 13-year-olds start tentatively experimenting with alcohol, usually within the family environment. This reflects a desire, especially in boys, to move on from child status.
- 14- to 15-year-olds prefer to drink outside the family environment and are more secretive, hiding their behaviour from their parents. This age group tends to drink to get drunk, with the aim of testing their limits and having fun.
- 16- to 17-year-olds, particularly 17-year-olds, move on from experimentation to seeing themselves as more responsible drinkers, with a belief that they know their own limits. They see their early excess use of alcohol as an inevitable part of growing up. The role of parents also shifts as there is a growing acceptance of drinking as part of normal adolescent activity and towards trust in the young person as a 'responsible' drinker.

Men aged 16-24 are the heaviest drinking age-group of the population whilst for young women, consumption reaches its peak in the late teenage years. This is sometimes referred to as a 'rite of passage' phase. It is widely accepted that consumption by both sexes declines with the formation of steady relationships, parenthood and financial responsibilities. However, changing social patterns such as women working longer before having children and divorce/separation means that people are tending to drink and socialise more outside the home.

What young people drink

There is evidence that young people increasingly favour higher strength alcoholic drinks – stronger brands of beer, cider and lager as well as spirits.

In 2008, among 11- to 15-year-olds in England who had drunk alcohol in the last week, boys were more likely to have drunk beer, lager or cider (88%), spirits (60%) or alcopops (53%). Girls were most likely to have drunk alcopops (69%), spirits (73%) or beer, lager or cider (55%).

In general, there are no significant differences between age groups in the types of drinks consumed. The

ALCOHOL CONCERN

exception is that younger pupils are more likely to drink shandy than older pupils.

Where young people drink

It is illegal for young people under the age of 18 to purchase alcohol and – with the exception of over-16s in certain circumstances – they cannot consume alcohol in licensed premises.

The law permits children and young people from the age of five to drink alcohol at home.

Among pupils who drink alcohol, the proportion who drink in pubs or bars has fallen from 13% in 1996 to 7% in 2008. There has been an increase over the same period in the proportion who usually drink at home or someone else's home (from 52% to 64%), at parties with friends (from 23% to 33%) or out of doors (from 21% in 1999 to 27% in 2008).

There is evidence of a trend towards increased unsupervised drinking by young people in open-air public places – in parks, at bus stops and in shopping areas. The proportion of 11- to 15-year-olds who drink on the street, in a park or somewhere else outside has increased from 21% in 1999 to 27% in 2008.

Recent studies and reviews have noted the pattern of excess unsupervised drinking in public places and the potential problems that this causes for young people in terms of risk of accidents, fights and assaults. For example, a study in the North West suggests that 40% of young people who drank outside in public had experienced alcohol-fuelled violence either as victims or perpetrators.

Obtaining alcohol

It is unusual for 11- to 15-year-olds to purchase alcohol from a shop or supermarket (11%) and from pubs or bars (6%). However, those who did try to purchase alcohol at a pub or a bar were successful (82%), as were those who tried to buy from a shop (73%). Older teenagers were more successful than younger ones: 89% of 15-year-olds successfully bought alcohol at a pub or club compared to 59% of 11- to 13-year-olds.

In 2008, 42% of pupils had obtained alcohol in the last four weeks, most commonly by being given it by friends (24%) or parents (22%) or asking someone else to buy it (18%).

A study of risky drinking among 15- and 16-year-olds in the North West of England found that among drinkers, 34.1% stated that they bought their own alcohol and these individuals were more likely to engage in risky behaviour, e.g. three times more likely to binge-drink once or more a week. Those who had alcohol provided to them by parents (48.5% of drinkers) were 1.64% times less likely to binge-drink each week and 1.28% times less likely to drink in public places.

First drinks

It is not easy to pinpoint the actual age when young people first try alcohol. Recall of this event is usually vague and as children get older, their definitions of a proper drink change.

More crucial than the first drink is the age young people start to drink unsupervised, signifying a shift to drinking with friends rather than parents, and in open spaces, clubs and pubs rather than at home. This is around the age of 14 and 15.

Coleman and Cater in their study of risky teenage drinking particularly noted that drunken episodes rather than consumption of alcohol 'appears to mark a crucial transition to repeated episodes of excessive drinking' and recommend a greater understanding of the 'process triggering the transition from first ever alcohol to first drunkenness'.

Why do young people drink?

Young people drink for much the same reasons as everyone else does – to have fun, to relax, to socialise and to feel more outgoing.

A study into underage risky drinking found that the young people interviewed cited three main reasons or motivations for drinking:

- Social facilitation – this was the most commonly reported explanation. This was linked to increased enjoyment and comfort in a social situation. It was also related to increased confidence in a social group and in sexual interaction.
- Individual benefits – the reasons for drunkenness were diverse but included escapism, getting a 'buzz' and having something to do.
- Social norms and influences – drinking and drinking to excess was seen as part of wider social norms and the accepted culture of heavy drinking, peer influence (including peer pressure) and for greater 'respect and image' among the social group. Although this study was specifically about risky drinking patterns, the authors found that the reasons given were similar to those given in studies of more general populations of young people.

March 2009

- The above information is an extract from the Alcohol Concern factsheet *Young people and alcohol* and is reprinted with permission. Visit www.alcoholconcern.org.uk for more information.

Your guide to alcohol units and measures

Information from Drinkaware.

How much are you drinking? We guide you through the confusing maze of labels, units and measures.

All drinks are not created equal

Discovering how much you are drinking sounds easy doesn't it? After all, you can drink a pint or a half pint of beer or cider, a single or a double measure of spirits or a glass of wine. And if you're counting your units it's easy – two units for a pint, and one unit for half a pint or a spirit and mixer…well, not quite. Unfortunately, the truth is more complicated than that.

How drunk you get depends on how much pure alcohol your drink contains. One way to calculate this is by counting units. The Government's recommended guidelines are up to two to three units of alcohol a day for a woman, and up to three to four for a man.

One unit is 10 ml of pure alcohol – the amount of alcohol the average adult can process within an hour. This means that if the average adult drinks a drink with one unit of alcohol in it, within an hour there should in theory be no alcohol left in their bloodstream, but that length of time could differ depending on a person's body size.

The alcohol content in drinks is also expressed as a percentage of the whole drink. If you take a look at the label of a bottle of wine or a can of lager you will see either a percentage, followed by the abbreviation 'ABV' which stands for alcohol by volume, or sometimes just the word 'vol'. So, wine that says '13 ABV' on its label contains 13% pure alcohol.

So how strong is your drink?

But the alcoholic content in the same types of drinks can vary a lot. Our tastes in alcohol have also changed. Warm, flat ale may have been our national drink 30 years ago, but imported lagers have gradually become more and more popular, and their alcohol strength can be quite a bit higher than ale. For example, some ales have an alcohol content of 3.5%. But stronger continental lagers can be 5% ABV, or even 6%.

This means that just one pint of strong lager can equate to more than three units of alcohol – almost your daily recommended guideline if you are a man. And research shows that we are drinking stronger drinks than ever before.

We also drink more wine than we used to, particularly wines from hot countries, including Australia, South America and South Africa. And whereas most wines are between 12 and 14% ABV, these robust, new world wines can be 17% ABV – or more.

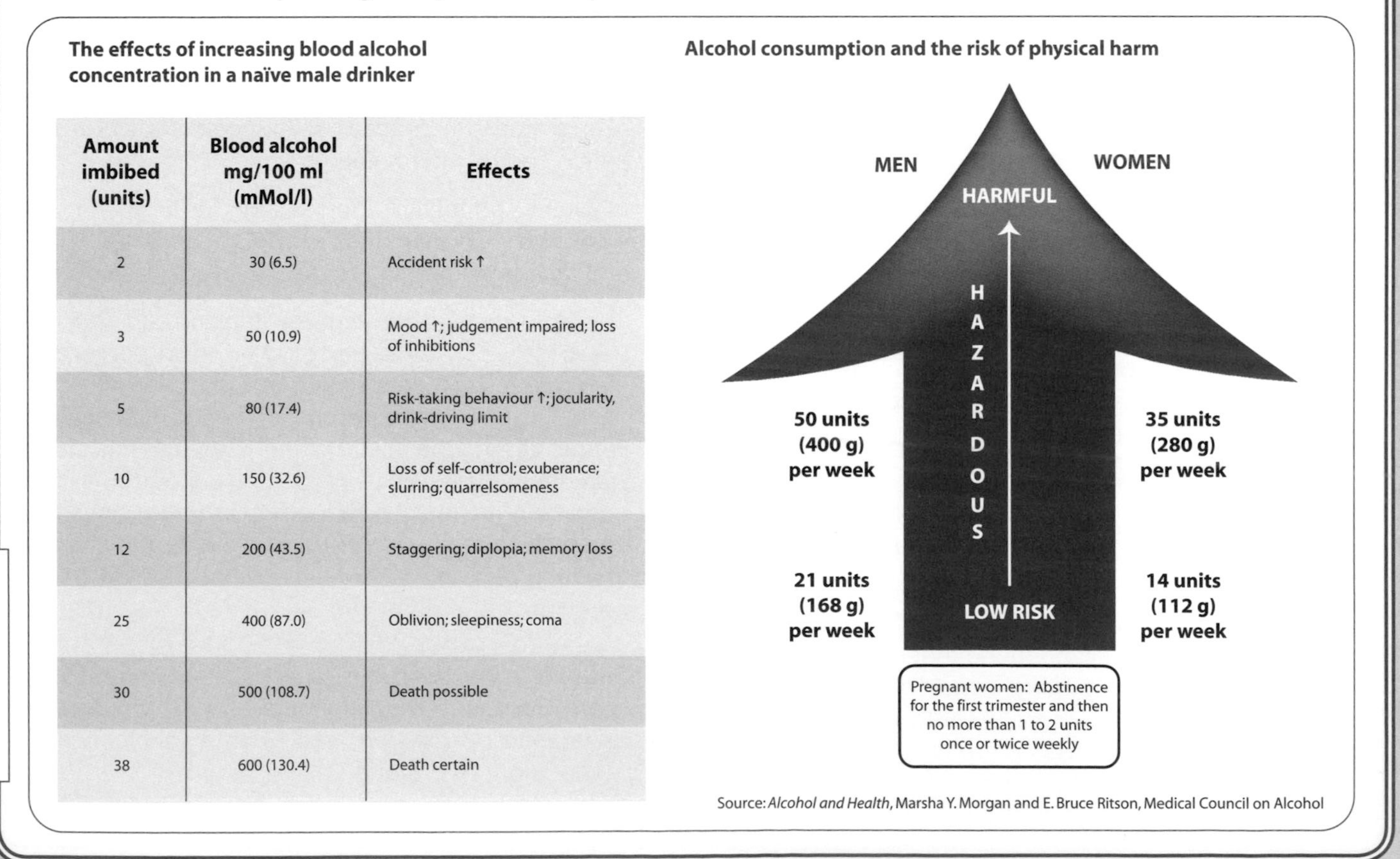

The effects of increasing blood alcohol concentration in a naïve male drinker

Amount imbibed (units)	Blood alcohol mg/100 ml (mMol/l)	Effects
2	30 (6.5)	Accident risk ↑
3	50 (10.9)	Mood ↑; judgement impaired; loss of inhibitions
5	80 (17.4)	Risk-taking behaviour ↑; jocularity, drink-driving limit
10	150 (32.6)	Loss of self-control; exuberance; slurring; quarrelsomeness
12	200 (43.5)	Staggering; diplopia; memory loss
25	400 (87.0)	Oblivion; sleepiness; coma
30	500 (108.7)	Death possible
38	600 (130.4)	Death certain

Source: *Alcohol and Health*, Marsha Y. Morgan and E. Bruce Ritson, Medical Council on Alcohol

Mintel, the market researchers, have found that the amount of pure 100% alcohol consumed by British drinkers has increased by 10% since 2000, despite the actual volume of alcohol consumed (in litres) remaining static throughout this period.

'It may be that the majority of consumers are not aware of ABV and don't even notice. So despite a greater societal concern with being healthy … by stealth we are drinking more pure alcohol than ever,' says Jonny Forsyth, a senior drinks analyst at Mintel.

Large or small?

And don't forget about size either.

Although spirits used to be commonly served in 25 ml measures, which are one unit of alcohol, many pubs and bars now serve them in 35 ml or 50 ml measures.

And if you ask for a glass of wine in a bar you'll be asked if you want a large or a small glass. You may opt for a large glass thinking you're getting a bargain, but a large measure is 250 ml – which is one-third of a bottle. This can be nearly three units of alcohol or more in just one glass. So if you have just two or three drinks, you could easily consume a whole bottle of wine – and almost three times your guideline daily units of alcohol – without even realising.

My drinking strategy:
+ + + = walk home
water
crisps
Mine:
+ + +
+ + = Stagger home

As part of its proposed Mandatory Sale of Alcohol Code, the Government plans, amongst other things, to make it compulsory for pubs and bars to offer their customers the choice of single measures and small wine glasses.

Strategies for lower alcohol drinking

Out and about

- Ask for a small glass of wine – as well as serving wine in smaller 175 ml glasses, some bars serve 125 ml glasses of wine – that can be one and a half units of alcohol.
- Drink spritzers if you like wine, or pints of shandy if you are a lager drinker. You will get a large drink, but one that contains less alcohol.
- Opt for half pints if you prefer higher strength lager or try lower strength beer – you really won't notice the difference.
- Alternate alcoholic drinks with soft drinks.
- Ask questions. If you are still uncertain about how much you are drinking, ask the bar staff. Do they pour doubles or singles? How big is their large glass of wine?

And at home …

If you're pouring your own drinks at home, it's easy to drink more than you would usually. Here are some tips to help you keep track of your intake.

- Measure spirits instead of free pouring them. Invest in some funky kitchen equipment – you can buy spirit measures and pourers in most good kitchenware shops – or online.
- Or use your imagination! Use an egg cup to pour measures, for example – check how much liquid it contains first by using a measuring jug.
- If you drink wine at home, pour small amounts into your glass. If you fill glasses to the rim, you'll be drinking more than you realise.
- Let guests pour their own drinks. If your half-full glass is constantly topped up, it's hard to keep track of how much you are drinking.

6 October 2009

- The above information is reprinted with kind permission from Drinkaware. Visit www.drinkaware.co.uk for more information.

DRINKAWARE

Effects of alcohol on your health

Information from Drinkaware.

From the second you take your first sip, alcohol starts affecting your body and mind. After one or two drinks you may start feeling more sociable, but drink too much and basic human functions, such as walking and talking become much harder. You might also start saying things you don't mean and behaving out of character. Some of alcohol's effects disappear overnight – while others can stay with you a lot longer, or indeed become permanent.

On the Drinkaware website you'll find useful clinically approved facts and information about the effects of alcohol on your life and lifestyle designed to help you make positive decisions about your drinking.

Diseases and cancers

Experts estimate alcohol is responsible for at least 33,000 deaths in the UK each year. While rates of liver disease are falling in the rest of Europe, they are rising in the UK. A 2006 *Lancet* study found that liver cirrhosis death rates are already around twice as high in Scotland as they are in other European countries.

Liver disease used to affect mainly drinkers in middle age, but now sufferers are getting younger. Up to one in three adults in the UK drinks enough alcohol to be at risk of developing alcohol-related liver disease.

Alcohol misuse is an important factor in a number of cancers, including liver cancer and mouth cancer, both of which are on the increase. Alcohol is second only to smoking as a risk factor for oral and digestive tract cancers.

Evidence suggests that this is because alcohol breaks down into a substance called acetaldehyde, which can bind to proteins in the mouth. This can trigger an inflammatory response from the body – in the most severe cases, cancerous cells can develop.

Chronic pancreatitis is another disease associated with heavy drinking. It's caused when your pancreas becomes inflamed and cells become damaged. Diabetes is a common side effect of chronic pancreatitis. There's evidence that heavy drinking can reduce the body's sensitivity to insulin, which can trigger type 2 diabetes.

While studies suggesting that alcohol can help heart disease often hit the headlines, the reality is that the jury's still out on the extent of any benefits. And it is clear that any benefits which there may be are limited to very low levels of consumption – probably no more than one unit of alcohol per day.

Mental health

Alcohol alters the brain's chemistry and increases the risk of depression. It is often associated with a range of mental health problems. A recent British survey found that people suffering from anxiety or depression were twice as likely to be heavy or problem drinkers.

Extreme levels of drinking (defined as more than 30 units per day for several weeks) can occasionally cause 'psychosis', a severe mental illness where hallucinations and delusions of persecution develop. Psychotic symptoms can also occur when very heavy drinkers suddenly stop drinking and develop a condition known as 'delirium tremens'.

Experts estimate that one in 17 people (6.4%) in Great Britain depend on alcohol to get through the day

Heavy drinking often leads to work and family problems, which in turn can lead to isolation and depression. For heavy drinkers who drink daily and become dependent on alcohol, there can be withdrawal symptoms (nervousness, tremors, palpitations) which resemble severe anxiety, and may even cause phobias, such as a fear of going out.

Appearance

If you're trying to watch your waistline, drinking too much alcohol can be disastrous! Research from the Department of Health reveals that a man drinking five pints a week consumes the same number of calories as someone getting through 221 doughnuts a year.

Drinking too much alcohol isn't great news for your skin either. As well as causing bloating and dark circles under your eyes, alcohol dries out your skin and can lead to wrinkles and premature ageing. If you drink heavily you may develop acne rosacea, a skin disorder that starts with a tendency to blush and flush easily and can progress to facial disfigurement, a condition known as rhinophyma.

Dependence

If you drink large quantities of alcohol on a regular basis you run the risk of becoming addicted. Experts estimate that one in 17 people (6.4%) in Great Britain depend on alcohol to get through the day. This can have serious effects on their families, friends and partners, as well as their mental health.

DRINKAWARE

Alcohol poisoning

Between 2007 and 2008 more than 30,000 people were admitted to hospital with alcohol poisoning. In the worst cases alcohol poisoning can cause lung damage (as you inhale your own vomit) and even lead to a heart attack.

Many traditional 'cures', such as drinking black coffee; just don't work – or even make things worse.

The morning after

If you've drunk heavily the night before, you'll almost certainly wake up with a hangover. Alcohol irritates the stomach, so heavy drinking can cause sickness and nausea and sometimes diarrhoea. Alcohol also has a dehydrating effect, which is one reason why excessive drinking can lead to a thumping headache the morning after.

Alcohol is a depressant, not a stimulant. This means that it slows down the brain and the central nervous system's processes. You may wonder what you did the night before, feel guilty, low or lethargic.

Women and alcohol

These days women are just as likely as men to make alcohol a major part of their social lives. The problem is that many women regularly drink more than the Government's daily recommended guidelines of two to three alcohol units, with around one in 14 drinking alcohol every day.

Women respond to alcohol differently from men, so the recommended levels are lower than for their male counterparts.

Recent research such as Oxford University's *Million Women Study* highlights the links between moderate drinking and increased risks of breast cancer.

19 April 2010

⇨ The above information is reprinted with kind permission from Drinkaware. Visit www.drinkaware.co.uk for more information.

Alcohol facts and trivia

***Random information that could save your life, or provide the winning answer on* Who Wants To Be A Millionaire?**

- ⇨ Alcohol features in almost a third of all UK divorce petitions, which means one or both partner's drinking habits have contributed to the bust up.
- ⇨ 13,000 violent incidents take place outside UK bars, pubs or licensed premises, every week. Most are alcohol-related.
- ⇨ Recent reports suggest that almost 50% of all British teenagers know how to buy bootlegged alcohol that has been smuggled into the country.
- ⇨ Alcohol robs your body of vitamin B complex, a vital group of nutrients, deficiency of which can cause skin damage, diarrhoea and depression.
- ⇨ Booze is believed to feature as a reason behind 25% of all school exclusions in the UK.
- ⇨ Women may find their reaction to booze is affected by hormone level changes that occur during the menstrual cycle.
- ⇨ An average 12% of males and 7% of females in the 16-19 age group show signs of alcohol dependency (source: Office of Population and Census Surveys).
- ⇨ Alcohol-related reasons and excuses are responsible for roughly 14 million lost working days in the UK every year.
- ⇨ Booze features in almost 40% of all domestic violence incidents in the UK.
- ⇨ Every Christmas in the UK, 10,000 people seek help for alcohol-related problems.

⇨ The above information is reprinted with kind permission from TheSite.org. Visit www.thesite.org for more information.

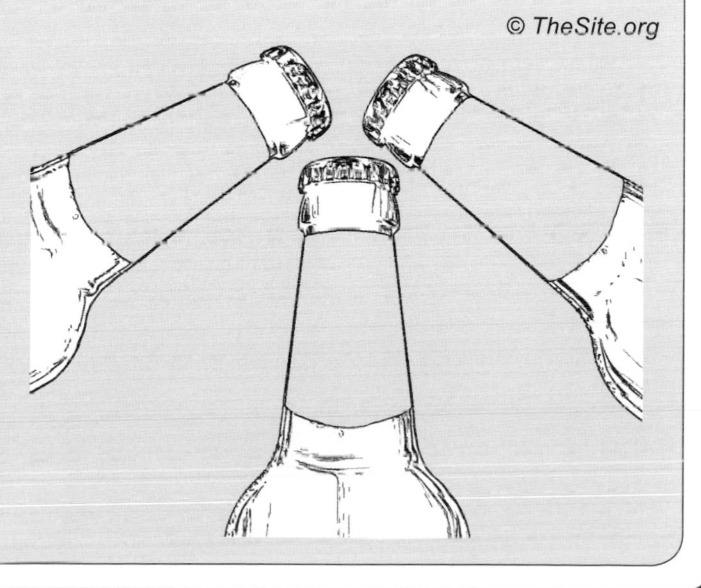

Drink spiking

Information from NHS Choices.

Introduction

Drink spiking is when mind-altering substances, such as drugs or alcohol, are added to your drink without you knowing. Mind-altering means that it may affect how you act, or how you behave with other people.

Who is at risk?

The people who are most at risk from drink spiking are those who regularly drink too much alcohol. For men, the recommended limit of alcohol is 21 units a week (three to four units a day), and for women the recommended limit is 14 units (two to three units a day). A unit of alcohol is equal to about half a pint of normal strength lager, a small glass of wine, or a pub measure (25 ml) of spirits.

NHS CHOICES

There are many reasons why someone might spike a drink, and it is not only females who could be targeted. The most common reasons are:

- for amusement.
- to be malicious (deliberately nasty).
- to carry out a sexual assault or rape.
- to carry out a physical assault.
- to carry out a theft.

Outlook

The symptoms of drink spiking will depend on whether alcohol, or another drug, has been used, how much of the substance was used, and how much alcohol you have already drunk. You will need to have your blood or urine tested by the police to confirm that your drink has been spiked with drugs.

Drink spiking is illegal, even if an attack or assault has not been carried out. It can result in a maximum punishment of ten years in prison for anyone who is found guilty of doing it. If an assault, rape, or robbery is also carried out, the sentence will be even higher.

Symptoms of drink spiking

If your drink has been spiked, your symptoms will depend on what drug has been used. The effect of any drug will depend on your body shape and size, your age, how much of the spiked drink you have consumed, and how much alcohol (if any) you have already drunk.

Any drug could be slipped into your drink without you knowing. Drugs can come in powder or liquid form, and may not have a taste or smell that you can identify as unusual. See the A-Z of drugs on the Frank website for more information about illegal substances and their effects.

Date rape drugs

The most common drugs that are used in drink spiking are often referred to as date rape drugs. This is because they make it harder for you to resist an assault. The most common date rape drugs are:

- alcohol.
- gamma-hydroxybutyrate (GHB) and gamma-butyrolactone (GBL).
- tranquilisers, most often benzodiazepines, including valium and rohypnol.
- ketamine.

These drugs are depressants which work by slowing down your nervous system, and dulling your responses and your instincts. In moderation, alcohol can help to relax you, and some date rape drugs are legally prescribed for anxiety and insomnia. However, when taken without knowing, these substances leave you vulnerable to danger.

Date rape drugs will affect your behaviour and the messages that you give out to other people. You will not be fully in control of yourself and someone could take advantage of you.

Any drug could be slipped into your drink without you knowing. Drugs can come in powder or liquid form, and may not have a taste or smell that you can identify as unusual

Date rape drugs can start to take effect within five minutes of being taken, or up to an hour after being taken. The symptoms for the above drugs, including alcohol, are quite similar, and will include some of the following:

- drowsiness or light-headedness.
- difficulty concentrating.
- feeling confused or disorientated, particularly after waking up (if you have been asleep).
- difficulty speaking, or slurring your words.
- loss of balance and finding it hard to move.
- lowered inhibitions.
- paranoia (a feeling of fear or distrust of others).
- amnesia (memory loss) or a 'black-out' of events (when you cannot remember large sections of your evening).
- temporary loss of body sensation (feeling like you are floating above your body, or having an 'out of body' experience).
- visual problems, particularly blurred vision.
- hallucinations (seeing, hearing, or touching things that are not really there).
- nausea and vomiting.
- unconsciousness.

All date rape drugs are particularly dangerous when they are mixed with alcohol because they combine to have a very powerful anaesthetic effect. This causes unconsciousness and, in more extreme cases, it can cause coma or even death.

How long the effects of the drugs last will depend on how much has been taken and how much alcohol, if any, you have drunk. The symptoms could last between three to seven hours, but if you pass out it will be hard to know the full effect. You may still feel some of the symptoms of a date rape drug after a night's sleep, particularly confusion, amnesia or nausea.

The most common date rape drugs are described in more detail below.

Alcohol

Alcohol is the most common date rape drug. It can be added to a soft (non-alcoholic) drink without you knowing, or double measures can be used instead of singles. If you have had a drink already, you may find it harder to tell how much alcohol you are consuming. The effects of alcohol will depend on how much you drink, and if you have been drinking already.

In large amounts, alcohol can be very dangerous, particularly if you pass out and vomit in your sleep. It takes your body one hour to process a unit of alcohol, so how long the effects last will depend on how many units of alcohol you have consumed.

Gamma-hydroxybutyrate and gamma-butyrolactone

Gamma-hydroxybutyrate (GHB) usually comes in the form of a slightly oily, colourless, liquid, and less often as a powder.

Gamma-butyrolactone (GBL) is a more basic form of GHB and another possible date rape drug. It comes in liquid form and is found in some household products. After entering the body, GBL changes into GHB.

Only a very small amount of GHB is needed in order to have an effect, and it can be dissolved easily into other liquids. GHB has an unpleasant taste, and a weak odour but, in very small doses, or if is mixed with a strong flavoured drink, you are unlikely to notice it.

Tranquilisers

Tranquilisers come in hundreds of different forms, but the most common are called benzodiazepines. You may hear of these as valium, rohypnol, roofies or benzos. They are sometimes legally prescribed to treat anxiety or insomnia. Tranquilisers work by slowing down your body, relieving tension, and making you feel very relaxed. They normally come as a tablet.

Ketamine

Ketamine, sometimes just called K, is a powerful anaesthetic that is used for both animals and humans.

NHS CHOICES

In its legal form it is a liquid, but illegally, it is normally a grainy white powder or a tablet. Ketamine can cause hallucinations (when you see or hear things that are not real) or it can create a feeling of your mind being separate from your body.

Preventing drink spiking

If your drink has been spiked, it is unlikely that you will be able to see, taste or smell any difference, so it is important that you try to prevent it from happening.

Follow the guidelines below to help lower the risk of having your drink spiked, and to help you to stay safe when you are out.

- keep your drink in your hand, and hold your thumb over the opening if you are drinking from a bottle.
- keep an eye on your friend's drinks.
- do not leave your drink unattended at any time, even while you are in the toilet.
- never accept a drink from anyone you do not know or trust.
- do not share or exchange drinks, or drink leftover drinks.
- when possible, drink from a bottle rather than a glass because it is more difficult to spike a drink in a bottle.
- stay away from situations that you do not feel comfortable with.
- if you go on a date with someone who you do not know, tell a friend or relative where you will be and what time you will be back.
- do not give away too much information to anyone you have just met, such as your address.
- do not show off expensive equipment, like mobile phones, or MP3 players, as you may attract unwanted attention.
- remember that if you have already been drinking, you will be more vulnerable because alcohol dulls your instincts and your awareness of danger.
- do not become complacent – always remember that it could happen to you.

Stopper devices and testing kits

Some pubs and clubs provide plastic stopper devices, such as lids to put on bottles, which can lower the risk of your drink being spiked.

Some also provide kits so that you can test your drink to see if it has been spiked. However, these may not test for every kind of drug and they are not always accurate. Tests that rely on changing the colour of your drink to indicate a positive result will not work if your drink is already quite dark – for example cola, blackcurrant cordial or red wine.

If you think that your drink has been tampered with, do not drink it and tell the management of the club or pub that you are in immediately.

16 September 2009

- Reproduced by kind permission of the Department of Health.

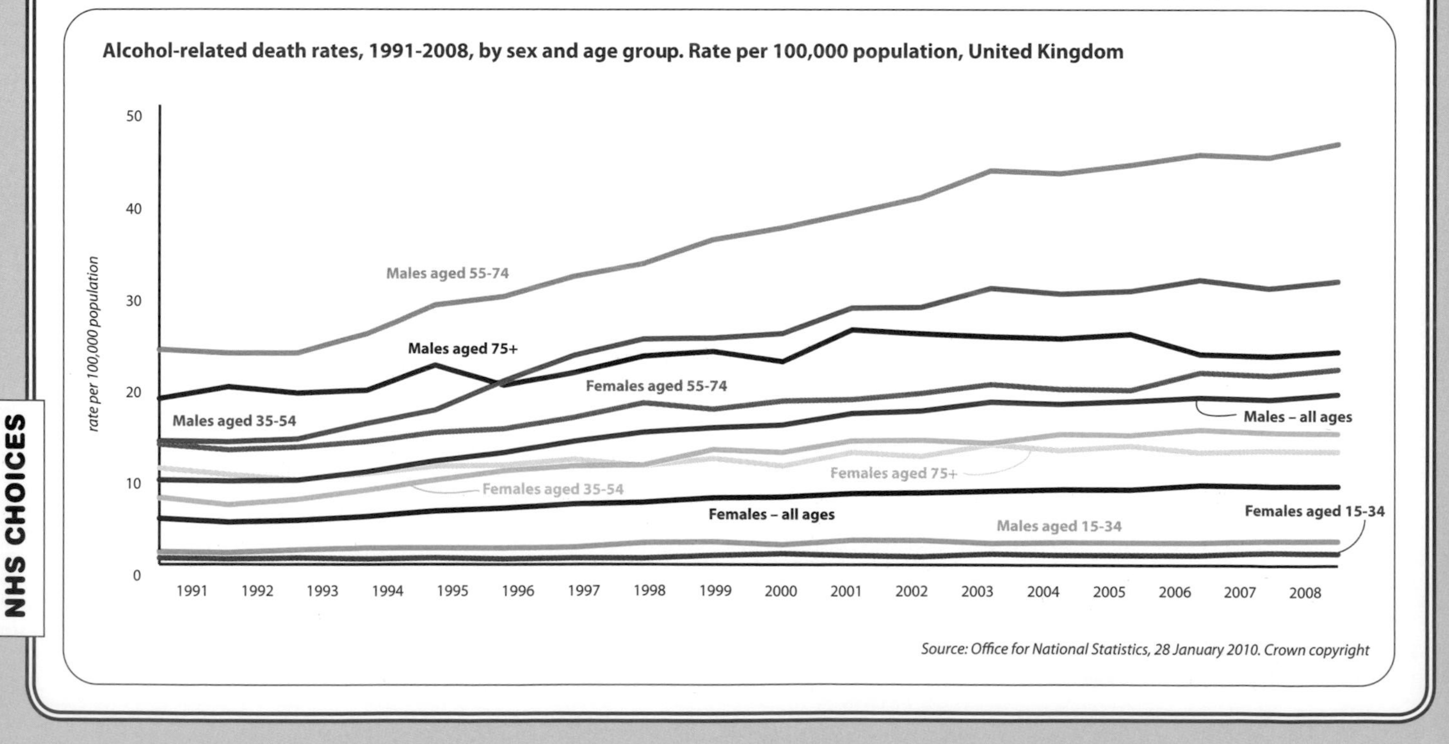

NHS CHOICES

Being teetotal

No more hangovers, puking in your hair or forgetting what you got up to the night before – welcome to the world of the teetotaller.

By Susie Wild

What is a teetotaller?

A teetotaller is somebody who abstains from drinking alcohol. They may have never tried it, they may have drunk it in the past, but they won't touch the stuff now. Nor will they eat it, so even vodka jelly is off the menu. Famous teetotallers include Chris Martin, Ewan McGregor, Jennifer Lopez, Tobey Maguire, Natalie Portman, and Rachel Stevens.

Why do people give up alcohol?

People choose to avoid alcohol for all sorts of reasons. Some teetotallers give up alcohol due to more serious problems, like alcohol dependency or addiction. Or it may be that they want to be healthier, get rid of their spots, or lose some weight. It could be that they want to save money, or they hate hangovers. Perhaps they don't like the feeling of being drunk or losing control, don't like the taste of alcohol, or have had bad drunken experiences.

Nick, 22, used to drink heavily from the age of 17. He gave up alcohol six months ago. 'I found myself recovering for half a week after drinking, which was playing havoc with my life. Now I feel healthier and more awake during the day. My skin has cleared up and I can actually remember nights out,' he says.

A teetotaller is somebody who abstains from drinking alcohol. They may have never tried it, they may have drunk it in the past, but they won't touch the stuff now

People may also choose to be teetotal for religious or spiritual reasons – most Hare Krishnas, Muslims, Scientologists, Sikhs, Seventh-day Adventists, Mormons, Brahmins and Bahá'ís are likely to be teetotal as part of the belief of their religion, but there are exceptions. Christians, including Methodists and Quakers, are also associated with teetotalism.

Alcohol and social situations

Culturally, most of our adult socialising revolves around pubs, clubs and parties – places where booze is easily available. Yet despite this fairly closed idea of having fun there are many other things you can do to have a good time without getting drunk. Honest!

Ellie, 25, hasn't had a drink for the past six months and says she has a much better time without it. She suffers from depression and found that alcohol made it much worse. 'Everyone has been really supportive. I meet a few people who think it's really sad that I can't "go out and enjoy myself", but they're usually avoiding looking at their own drinking habits,' she explains. 'It totally baffles them that I've given up drinking completely. I don't really miss drinking – I'm just happy to know I'll never have another hangover, be sick in my hair, or wake up next to a stranger.'

When you go out you may order alcohol more out of habit than a desire to drink it. Your friends may give you a hard time for ordering a soft drink instead; people don't like to drink booze alone, It can also unnerve them that you want to remain sober.

THESITE.ORG

How to deal with peer pressure to drink

If you're not driving, ill, or pregnant, friends may pile on the pressure for you to join in. 'When I go out everyone tries to persuade me to drink with them,' says Nick. 'I get negative comments from friends because they think I can't enjoy myself as much as I would if I was drinking. I avoid alcohol by distracting myself with other hobbies, or just socialising at home.'

When you go out you may order alcohol more out of habit than a desire to drink it. Your friends may give you a hard time for ordering a soft drink instead; people don't like to drink booze alone

Ultimately, alcohol is not an essential part of a night out. You can still have fun at parties and clubs but you'll have a clearer idea of who you're pulling, and how you're getting home. Just make sure you're not lumbered with all the responsibilities of the night, such as being the designated driver. Remind your friends that you're out to have fun, too, and if they're still giving you grief, try some of these tactics:

- ⇨ Tell them the truth – that you don't drink or don't want to drink.
- ⇨ Skip out of rounds and avoid telling people that you're not drinking alcohol.
- ⇨ You may decide to avoid socialising with people who are drinking, or in places where alcohol is available.
- ⇨ If your friends can't respect your decision to stay sober, you may decide it's time to find other like-minded people to hang out with.

Alternatives to booze

There are plenty of places that serve good non-alcoholic drinks and where the nature of the entertainment is not focused on getting drunk. The trick is to think about what appeals to you and what makes a good night out. Here are some suggestions:

- ⇨ Art centres, late night café-bars and independent cinemas are good alternatives to pubs and clubs.
- ⇨ Ten-pin bowling and bingo have had a resurgence in popularity, as have ironic kitsch days of afternoon tea and craft.
- ⇨ Evening classes, social groups and art and sport clubs can offer ways to meet people and socialise where alcohol needn't even crop up.
- ⇨ Doing the pub quiz, having Sunday lunch or playing pool can switch the focus of trips to the pub away from simply getting smashed.

⇨ The above information is reprinted with kind permission from TheSite.org. Visit www.thesite.org for more information.

Respondents were asked: 'Which of these apply to you? Please choose one only.'

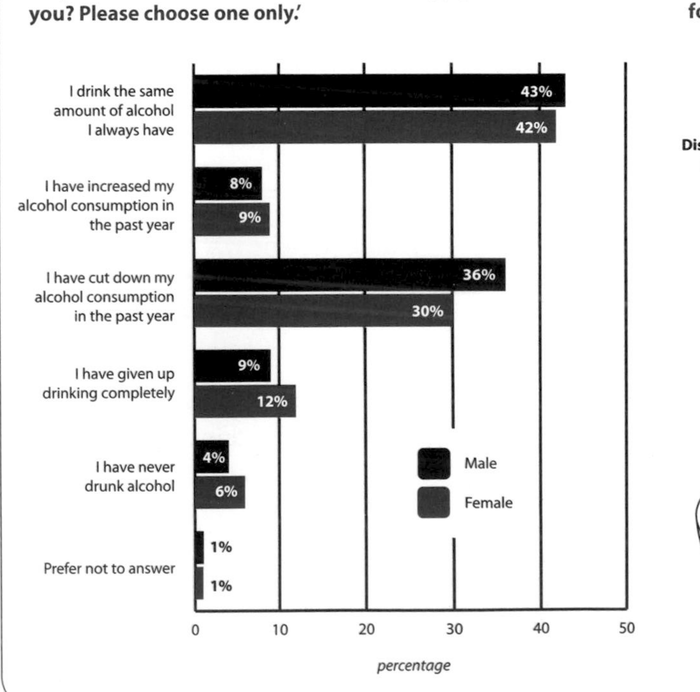

Respondents were asked: 'Do you agree or disagree with the following statements?'

It doesn't hurt to have a drink from time to time

Response	%
Agree	95%
Disagree	2%
Neither/not applicable	3%

Alcohol is bad for you

Response	%
Agree	44%
Disagree	37%
Neither/not applicable	18%

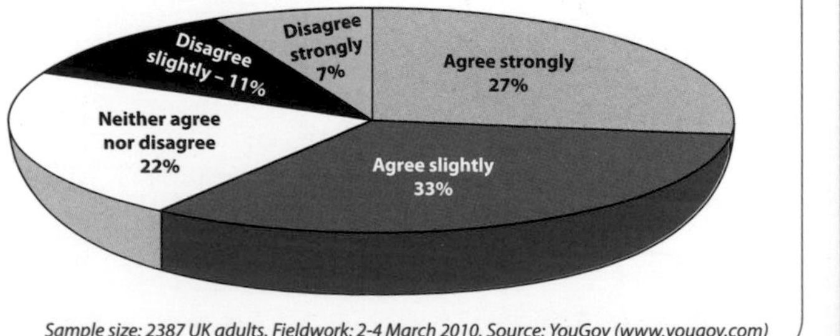

Sample size: 2387 UK adults. Fieldwork: 2-4 March 2010. Source: YouGov (www.yougov.com)

THESITE.ORG

Alcohol deaths

UK rates increase in 2008.

The number of alcohol-related deaths in the United Kingdom has consistently increased since the early 1990s, rising from the lowest figure of 4,023 (6.7 per 100,000) in 1992 to the highest of 9,031 (13.6 per 100,000) in 2008. Although figures in recent years suggested that the trend was levelling out, alcohol-related deaths in males increased further in 2008. Female rates have remained stable.

There are more alcohol-related deaths in men than in women. The rate of male deaths has more than doubled over the period from 9.1 per 100,000 in 1991 to 18.7 per 100,000 in 2008. There have been steadier increases in female rates, rising from 5.0 per 100,000 in 1991 to 8.7 in 2008, less than half the rate for males. In 2008, males accounted for approximately two-thirds of the total number of alcohol-related deaths. There were 5,999 deaths in men and 3,032 in women.

The trends differ according to age. For both males and females, the lowest rates across the period were in those aged 15-34. In 2008, the rates were 2.9 per 100,000 and 1.3 per 100,000, respectively. The highest rates have occurred in persons aged 55-74. In men, the rate has increased from 23.0 per 100,000 in 1992 and 1993 to 45.8 per 100,000 in 2008, the highest rate recorded across all age groups. Alcohol-related deaths in all age groups increased in 2008 compared with figures in 2007.

Alcohol-related death rates among females have been consistently lower than rates for males and trends demonstrate a broadly similar pattern across different age groups. As for men, the highest rates for women during the 1991-2008 period were in those aged 55-74. In 2008, the rate for this group peaked at 21.5 per 100,000. Rates in all other age groups decreased slightly and were lowest in women aged 15-34 at 1.3 per 100,000.

Source

Office for National Statistics, General Register Office for Scotland, Northern Ireland Statistics and Research Agency.

Notes

The ONS definition of alcohol-related deaths (which includes causes regarded as most directly due to alcohol consumption) was revised in 2006. Details can be found via 'Alcohol-related deaths in the UK' on the ONS website. ONS has agreed with the GROS and NISRA that this definition will be used to report alcohol-related deaths for the UK.

The introduction of the Tenth Revision of the International Classification of Diseases (ICD-10) for coding cause of death means that data following its implementation are not completely comparable with earlier years. Mortality data for England and Wales show that the introduction of ICD-10 resulted in a difference in the number of alcohol-related deaths below one per cent.

Rates are based on deaths registered in each calendar year and are directly age-standardised using the European Standard Population.

28 January 2010

⇨ The above information is reprinted with kind permission from the Office for National Statistics. Visit www.statistics.gov.uk for more information.

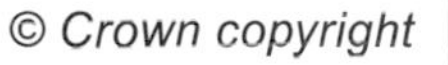

Number of alcohol-related deaths, 1991-2008, United Kingdom

Source: Office for National Statistics, 28 January 2010. Crown copyright

OFFICE FOR NATIONAL STATISTICS

Alcohol and sensible drinking

Information from Patient UK.

What are the recommended safe limits of alcohol drinking?

- Men should drink no more than 21 units of alcohol per week (and no more than four units in any one day).
- Women should drink no more than 14 units of alcohol per week (and no more than three units in any one day).
- The exact amount that is safe for pregnant women is not known. Therefore, advice from the Department of Health is that pregnant women and women trying to become pregnant should not drink at all. If you do choose to drink when you are pregnant then limit it to one or two units, once or twice a week. And never get drunk.

In general, the more you drink above the safe limits, the more harmful alcohol is likely to be. And remember, binge drinking can be harmful even though the weekly total may not seem too high. For example, if you only drink once or twice a week, but when you do you drink four to five pints of beer each time, or a bottle of wine each time, then this is a risk to your health. Also, even one or two units can be dangerous if you drive, operate machinery, or take some types of medication.

What is a unit of alcohol?

One unit of alcohol is 10 ml (1 cl) by volume, or 8 g by weight, of pure alcohol. For example:

- One unit of alcohol is about equal to:
  - Half a pint of ordinary strength beer, lager, or cider (3-4% alcohol by volume); or
  - A small pub measure (25 ml) of spirits (40% alcohol by volume); or
  - A standard pub measure (50 ml) of fortified wine such as sherry or port (20% alcohol by volume).
- There are one and a half units of alcohol in:
  - A small glass (125 ml) of ordinary strength wine (12% alcohol by volume); or
  - A standard pub measure (35 ml) of spirits (40% alcohol by volume).

But remember, many wines and beers are stronger than the more traditional 'ordinary' strengths. A more accurate way of calculating units is as follows. The percentage alcohol by volume (% ABV) of a drink equals the number of units in one litre of that drink. For example:

- Strong beer at 6% ABV has six units in one litre. If you drink half a litre (500 ml) – just under a pint – then you have had three units.
- Wine at 14% ABV has 14 units in one litre. If you drink a quarter of a litre (250 ml) – two small glasses – then you have had three and a half units.

Some other examples

Three pints of beer, three times per week, is at least 18-20 units per week. That is nearly the upper weekly safe limit for a man. However, each drinking session of three pints is at least six units, which is more than the safe limit advised for any one day. Another example: a 750 ml bottle of 12% wine contains nine units. If you drink two bottles of 12% wine over a week, that is 18 units. This is above the upper safe limit for a woman.

Isn't alcohol good for you?

For men over 40 and for women past the menopause, it is thought that drinking a small amount of alcohol (one to two units per day) helps to protect against heart disease and stroke.

Do you know how much you are drinking?

When asked 'How much do you drink?' many people give a much lower figure than the true amount. It is not that people usually lie about this, but it is easy to not realise

PATIENT UK

your true alcohol intake. To give an honest answer to this question, try making a drinking diary for a couple of weeks or so. Jot down every drink that you have. Remember, it is a pub measure of spirits that equals one unit. A home measure is often a double.

If you are drinking more than the safe limits, you should aim to cut down your drinking.

What are the problems with drinking too much alcohol?

Health risks

About one in three men, and about one in seven women, drink more than the safe levels. Many people who drink heavily are not 'addicted' to alcohol, and are not 'alcoholics'. To stop or reduce alcohol would not be a problem if there was the will to do so. However, for various reasons, many people have got into a habit of drinking regularly and heavily. But, drinking heavily is a serious health risk.

If you drink heavily you have an increased risk of developing:

- Hepatitis (inflammation of the liver);
- Cirrhosis (scarring of the liver). Up to three in ten long-term heavy drinkers develop cirrhosis;
- Stomach disorders;
- Pancreatitis (severe inflammation of the pancreas);
- Mental health problems including depression, anxiety, and various other problems;
- Sexual difficulties such as impotence;
- Muscle and heart muscle disease;
- High blood pressure;
- Damage to nervous tissue;
- Accidents – drinking alcohol is associated with a much increased risk of accidents. In particular, injury and death from fire and car crashes. About one in seven road deaths are caused by drinking alcohol;
- Some cancers (mouth, gullet, liver, colon and breast);
- Obesity (alcohol has many calories);
- Damage to an unborn baby in pregnant women;
- Alcohol dependence (addiction).

In the UK about 33,000 deaths a year are related to drinking alcohol, a quarter due to accidents.

Alcohol dependence

If you are 'alcohol dependent' you have a strong desire for alcohol and have great difficulty in controlling your drinking. In addition, your body is used to lots of alcohol. Therefore, you may develop withdrawal symptoms three to eight hours after your last drink as the effect of the alcohol wears off. So, even if you want to stop drinking, it is often difficult because of withdrawal symptoms.

Withdrawal symptoms include: feeling sick, trembling, sweating, craving for alcohol, and feeling unwell. As a result, you may drink regularly to prevent withdrawal symptoms.

The severity of dependence can vary. It can develop gradually and become more severe. You may be developing alcohol dependence if you:

- need a drink every day.
- drink alone often.
- need a drink to stop trembling (the shakes).
- drink early, or first thing in the morning (to avoid withdrawal symptoms).
- often have a strong desire to drink alcohol.
- spend a lot of you time in activities where alcohol is available; for example, if you spend a lot of time at the social club or pub.
- neglect other interests or pleasures because of alcohol drinking.

Alcohol drinking and problems to others

Heavy alcohol drinking in one person often seriously damages others. Many families have become severely affected by one member becoming a problem drinker. Emotional and financial problems often occur in such families. It is estimated that three in ten divorces, four in ten cases of domestic violence, and two in ten cases of child abuse are alcohol-related. Often the problem drinker denies or refuses to accept that the root cause is alcohol.

Some common myths about drinking alcohol

Myth – 'Coffee will sober me up'

Caffeine in coffee is a stimulant so you might feel more alert, but it won't make you sober.

Myth – 'I'll be fine in the morning'

Alcohol is broken down by the liver. A healthy liver can get rid of about one unit of alcohol an hour. Sleep will not speed up the rate at which the liver works. Just because you have a night's sleep does not necessarily mean you will be sober in the morning. It depends on how much you drank the night before.

PATIENT UK

Myth – 'Alcohol keeps me alert'

Alcohol can make you think that you are more alert, but it actually has a depressant effect which slows down your reflexes.

Myth – 'Beer will make me less drunk than spirits'

Half a pint of beer contains the same amount of alcohol as a single measure of spirits.

Myth – 'I'll be fine if I drink plenty of water before I go to bed'

This can reduce hangover symptoms by helping to prevent dehydration. But it wont make you any less drunk, or protect your liver or other organs from the damaging effect of alcohol.

In the UK about 33,000 deaths a year are related to drinking alcohol, a quarter due to accidents

Myth – 'The recommended safe limits are too low'

They are based on good research which has identified the level above which problems start to arise. For example, if a man drinks five units each day (not greatly over the recommended limit) then, on average, he doubles his risk of developing liver disease, raised blood pressure, some cancers, and of having a violent death.

Myth – 'Most people drink more than the recommended limits'

Studies show that about one in three men, and about one in seven women, drink more than the weekly recommended levels. So, if you drink heavily, it might be what your friends do, but it is not what most people do, and you are putting yourself and others at risk.

Myth – 'It's none of my business if a friend is drinking too much'

This is a matter of opinion. Some people would say that if you are a real friend, it really is your business. You may be the one person who can persuade your friend to accept that they have a problem, and to seek help if necessary.

Tackling the problem of heavy drinking

Once they know the facts, many people can quite easily revert back to sensible drinking if they are drinking above the safe limits. If you are trying to cut down, some tips which may help include:

- Consider drinking low alcohol beers, or at least do not drink 'strong' beers or lagers.
- Try pacing the rate of drinking. Perhaps alternate soft drinks with alcoholic drinks.
- If you eat when you drink, you may drink less.
- It may be worth reviewing your entire social routine. For example, consider:
  - cutting back on types of social activity which involve drinking;
  - trying different social activities where drinking is not a part;
  - reducing the number of days in the week where you go out to drink;
  - going out to the pub or club later in the evening.
- Try to resist any pressure from people who may encourage you to drink more than you really want to.

About one in three men, and about one in seven women, drink more than the weekly recommended levels

The problem of denial

Some people who are heavy drinkers, or who are alcohol dependent, deny that there is a problem to themselves. The sort of thoughts that people deceive themselves with include: 'I can cope', 'I'm only drinking what all my mates drink', 'I can stop anytime'.

Coming to terms with the fact that you may have a problem, and seeking help when needed, is often the biggest step to sorting the problem.

Do you need help?

Help and treatment is available if you find that you cannot cut down your drinking to safe limits. Counselling and support from a doctor, nurse, or counsellor is often all that is needed. A 'detoxification' treatment may be advised if you are alcohol dependent. Referral for specialist help may be best for some people.

If you feel that you, or a relative or friend, needs help about alcohol then see your doctor or practice nurse. Or, contact one of the agencies listed on the Patient UK website.

Reviewed: 12 January 2009

- The above information is reprinted with kind permission from Patient UK. Visit www.patient.co.uk for more information or to view references for this piece.

PATIENT UK

'Social drinking': the hidden risks

If you think it's only alcoholics and binge drinkers who are putting their health at risk, think again.

The NHS recommends:

- Men should not regularly drink more than three to four units of alcohol a day.
- Women should not regularly drink more than two to three units a day.

'Regularly' means drinking this amount every day or most days of the week.

Many of those who see themselves as 'social drinkers' are actually at risk of developing long-term health conditions because of the amount they drink on a regular basis.

Most drinkers are unaware that regularly drinking more than is advised by the NHS can lead to a wide range of long-term health problems, including cancers, strokes and heart attacks.

For a woman, simply having a large glass (250 ml) of 12% wine (three units) every day, or a man drinking two pints of 4% lager (4.6 units), can push you above the recommended limits

More than 55% of people questioned in a YouGov poll thought that alcohol only damaged your health if you regularly get drunk or binge drink.

The 2010 survey of 2,000 adults also found that 83% believed that regularly drinking more than is advised by the NHS didn't put their long-term health at risk.

The survey suggests there are possibly 7.5 million people who are unaware of the damage that their drinking could be causing.

Unseen damage

For a woman, simply having a large glass (250 ml) of 12% wine (three units) every day, or a man drinking two pints of 4% lager (4.6 units), can push you above the recommended limits.

Men who regularly drink more than two pints of strong (5.2%) lager, which is more than six units, every day:

- are more than three times more likely to get mouth cancer.
- could be three times more likely to have a stroke.

Women who regularly drink two large glasses of 13% wine (6.5 units) or more a day:

- are twice as likely to have high blood pressure.
- are 50% more likely to get breast cancer.

Over the limit

More than nine million people in England drink more than the recommended daily amount. More than 9,000 people in the UK die from alcohol-related causes each year. About 20% of these deaths are from cancer, 15% from cardiovascular illnesses, such as heart disease and stroke, and 13% are from liver disease.

While binge-drinking is usually associated with young adults, it's typically older people who drink more than the recommended levels for regular drinking, and who may consider themselves 'social drinkers', who will suffer longer term alcohol-related illness or death.

Professor Nigel Heaton is a liver transplant consultant at King's College Hospital, London. He says people who believe that drinking above recommended levels is just normal 'social drinking' are raising their risk of developing alcohol-related illnesses.

'Some people think it's natural to have a bottle of wine a night,' he says. 'It seems respectable because you're drinking with food and it's not associated with any drunken behaviour or even feeling drunk.

'But if it happens regularly, you may have problems later on. Most of us believe that people with alcoholic liver disease are alcoholics. We often think, "I'm not an alcoholic so I can't get liver disease."

'You may not be an alcoholic, but if the overall amount of alcohol you drink regularly exceeds recommended limits, it may still cause serious harm.'

1 March 2010

- Reproduced by kind permission of the Department of Health.

NHS CHOICES

Effects of alcohol

Brain damage

Binge drinking can cause blackouts, memory loss and anxiety. Long-term drinking can result in permanent brain damage, serious mental health problems and alcohol dependence or alcoholism. Young people's brains are particularly vulnerable because the brain is still developing during their teenage years. Alcohol can damage parts of the brain, affecting behaviour and the ability to learn and remember.

Cancers

Drinking alcohol is the second biggest risk factor for cancers of the mouth and throat (smoking being the first). People who develop cirrhosis of the liver (often caused by too much alcohol) can develop liver cancer.

Heart and circulation

Alcohol can cause high blood pressure (hypertension) increasing the risk of having a heart attack or stroke. It also weakens heart muscles, which can affect lungs, liver, brain and other body systems and can cause heart failure. Binge drinking and drinking heavily over longer periods can cause the heart to beat irregularly (arrhythmia) and has been linked to cases of sudden death.

Lungs

People who drink a lot of alcohol have more lung infections and can be more likely to get pneumonia and for their lungs to collapse. When a person vomits as a result of drinking alcohol they may choke if vomit gets sucked into their lungs.

Liver

Drinking too much alcohol initially causes fat deposits to develop in the liver. With continued excessive drinking the liver may become inflamed, resulting in alcoholic hepatitis which can result in liver failure and death. Excessive alcohol can permanently scar and damage the liver, resulting in liver cirrhosis and an increased risk of liver cancer.

Stomach

Drinking above recommended limits can lead to stomach ulcers, internal bleeding and cancer. Alcohol can cause the stomach to become inflamed (gastritis), which can prevent food from being absorbed and increase the risk of cancer.

Pancreas

Heavy or prolonged use of alcohol can cause inflammation of the pancreas, which can be very painful, causing vomiting, fever and weight loss, and can be fatal.

Intestine

Heavy drinking may result in ulcers and cancer of the colon. It also affects your body's ability to absorb nutrients and vitamins.

Kidneys

Heavy drinking can increase your risk of developing high blood pressure – a leading cause of chronic kidney disease.

Fertility

In men: impotence (lowered libido/sex drive) and infertility.

In women: infertility. Drinking alcohol when pregnant can seriously damage the development of the unborn baby.

Bones

Alcohol interferes with the body's ability to absorb calcium. As a result, your bones become weak and thin (osteoporosis).

Weight gain

Alcohol is high in calories. Weight for weight, the alcohol in a drink contains almost as many calories as fat. The average bottle of wine contains 600 calories while four pints of average strength lager contains 640.

Skin

Alcohol dehydrates your body and your skin; it also widens blood vessels causing your skin to look red or blotchy.

Sexual health

Binge drinking makes you lose your inhibitions and affects your judgement. This might make you less likely to use a condom, increasing your risk of getting a sexually transmitted infection such as chlamydia, HIV or hepatitis or result in an unplanned pregnancy.

Mental health

People may think alcohol helps them to cope with difficult situations and emotions, to reduce stress or relieve anxiety, but alcohol is in fact associated with a range of mental health problems including depression, anxiety, risk-taking behaviour, personality disorders and schizophrenia.

Alcohol has also been linked to suicide. The Mental Health Foundation* reports that:

- 65% of suicides have been linked to excessive drinking.
- 70% of men who kill themselves have drunk alcohol before doing so.
- almost one-third of suicides among young people take place while the person is intoxicated.

Excessive drinking can disrupt normal sleeping patterns, resulting in insomnia and a lack of restful sleep which can contribute to stress and anxiety.

**Mental Health Foundation.* Cheers! Understanding the relationship between alcohol and mental health. *London: Mental Health Foundation, 2006.*

⇨ Reproduced with permission from the Public Health Agency. Visit www.knowyourlimits.info for more.

PUBLIC HEALTH AGENCY

Young people and alcohol – what are the risks?

The effects of alcohol on young people are not the same as they are on adults. While alcohol misuse can present health risks and cause careless behaviour in all age groups, it is even more dangerous for young people. Find out how alcohol can affect young people's health and behaviour.

Health risks

Because young people's bodies are still growing, alcohol can interfere with their development. This makes young people particularly vulnerable to the long-term damage caused by alcohol. This damage can include:

- cancer of the mouth and throat.
- sexual and mental health problems.
- liver cirrhosis and heart disease.

Research also suggests that drinking alcohol in adolescence can harm the development of the brain.

Young people might think that any damage to their health caused by drinking lies so far in the future that it's not worth worrying about. However, there has been a sharp increase in the number of people in their twenties dying from liver disease as a result of drinking heavily in their teens.

Young people who drink are also much more likely to be involved in an accident and end up in hospital.

If it's twins then maybe I should call them Bacardi and Coke?

Risky behaviour – sex

Drinking alcohol lowers people's inhibitions, and makes them more likely to do things they would otherwise not do. Young people are particularly at risk because, at their stage of life, they are still testing the boundaries of what is acceptable behaviour.

One in five girls (and one in ten boys) aged 14 to 15 goes further than they wanted to in a sexual experience after drinking alcohol. In the most serious cases, alcohol could lead to them becoming the victim of a sexual assault.

Unsafe sex and unwanted pregnancy

If young people drink alcohol, they are more likely to be reckless and not use contraception if they have sex. Almost one in ten boys and around one in eight girls aged 15 to 16 have unsafe sex after drinking alcohol. This puts them at risk of sexual infections and unwanted pregnancy.

Research shows that a girl who drinks alcohol is more than twice as likely to have an unwanted pregnancy as a girl who doesn't drink.

Antisocial behaviour

Alcohol interferes with the way people think and makes them far more likely to act carelessly. If young people drink alcohol, they are more likely to end up in dangerous situations.

For example, they are more likely to climb walls or other heights and fall off. Or they might verbally abuse someone who could hit them. They are also more likely to become aggressive themselves and throw a punch.

Four out of ten secondary school-age children have been involved in some form of violence because of alcohol. This could mean they have been beaten up or robbed after they've been drinking, or have assaulted someone themselves.

Getting into trouble with the police

If a child or young person drinks alcohol, then they are more likely to get into trouble with the police. Every year, more than 10,000 fines for being drunk and disorderly are issued to young people aged 16 to 19.

DIRECTGOV

Children as young as 12 are being charged with criminal damage to other people's property as a result of drinking.

Criminal behaviour

Young people who get drunk at least once a month are twice as likely to commit a criminal offence as those who don't. More than one in three teenagers who drink alcohol at least once a week have committed violent offences such as robbery or assault.

Young people who get involved with crime are also likely to end up with a criminal record. This can damage their prospects for the rest of their life. Having a criminal record can exclude people from some jobs and, for some offences, prevent them from travelling abroad.

Failing to meet potential at school

When young people drink, it takes longer for the alcohol to get out of their system than it does in adults. So if young people drink alcohol on a night before school, then they can do less well in lessons the next day.

Young people who regularly drink alcohol are twice as likely to miss school and get poor grades as those who don't. Almost half of young people excluded from school are regular drinkers.

⇨ The above information is reprinted with kind permission from Directgov. Visit www.direct.gov.uk for more information.

The 24-hour drinking laws

Pubs and bars can now apply to stay open 24/7. Will this lead to a continental-style café culture or puke-splattered streets and cat fights round the clock? We look at the basics surrounding the law.

The licensing law allows pubs, clubs, bars, supermarkets and service stations in England, Scotland and Wales to apply for longer opening licenses. Critics of the law say the extension of the licensing hours encourages drunken goings-on in the streets.

The Government and supporters of the new law believe this is a good strategy to reduce booze-related crime and anti-social behaviour. They say the changes reduce the problem of post-pub brawling associated with hordes of people leaving different venues at the same fixed closing time.

What does the law say?

There are flexible opening hours for bars and pubs, with the potential to open 24 hours a day, seven days a week.

Venues are able to apply for a licence, but have to inform the local community and police to give them 21 days to object.

Any pubs that become disorderly as a result of the new opening hours will be given two months to sort out the situation or they will be billed for the extra policing costs.

The reality

At the time when the change in the law took effect, just over 60,000 outlets could sell alcohol for longer. But only about 1,000 outlets were granted the 24-hour license. Fewer than 400 of these are pubs and clubs, and the rest are supermarkets and service stations.

Some of the outlets that have received the 24-hour license may not even use them. A survey carried out by The British Beer & Pub Association has shown that the only days most venues will extend their licenses are Thursday, Friday and Saturday, with the majority expecting to close at 1 or 2am.

⇨ The above information is reprinted with kind permission from TheSite.org. Visit www.thesite.org for more information.

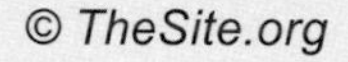

Bars with 24-hour drinking face 'law and order' levy

Late-night bars and pubs may be forced to pay a 'law and order' levy to help meet the costs of alcohol-fuelled disorder.

By Richard Edwards

Under plans to overhaul Labour's 24-hour drinking laws, the Government wants premises that stay open after 11pm to help pay for the cost of drunken violence on the streets.

Town halls will be given the power to charge them additional fees for late-night licences, with the amount likely to be graded on the establishment's popularity.

The advent of round-the-clock licensing, in 2005, has led to serious concerns that it has helped cause an increase in violence and alcohol abuse rather than to the sophisticated 'café culture' that Labour claimed.

The Government wants premises that stay open after 11pm to help pay for the cost of drunken violence on the streets

Officials are looking at changing the wording on licence applications so that pubs and clubs wanting to extend their hours will have to prove after-hours drinking offers a tangible 'benefit' to the local community.

Last month, in her first major speech as Home Secretary, Theresa May told the Police Federation annual conference that licensing laws were to undergo a 'complete review' after all the predicted problems had become a reality.

It is believed Mrs May wants alcohol to be considered a key law-and-order issue, and is attempting to have responsibility for licensing moved to the Home Office from the Department for Culture, Media and Sport.

The Association of Chief Police Officers and the Police Federation union have pressed for a hard-line stance on binge drinking.

Police chiefs claim that disorder related to alcohol is one of the biggest challenges facing forces.

Almost half of all violent crime victims report that their attacker was under the influence of alcohol.

In December, figures indicated that alcohol-related crime and disorder were costing every household almost £600 a year. The country was facing a total bill of about £13 billion to cope with the effects of drunken rowdiness and offending, including policing and health care.

Mrs May, speaking to rank and file officers last month, said: 'We are going to look at the licensing laws.

'I was in opposition when the new laws were introduced. I argued against them precisely because of the problems we have seen.

'I argued that those were the sorts of problems that would come about but I was told we would have a café culture.'

The National Institute for Health and Clinical Excellence (NICE) has called for a minimum price for alcohol to help tackle Britain's drink problem.

Civil servants are devising a formula to make the cost of alcohol more expensive in a bid to deter young drinkers. Stores will be banned from selling alcohol below the combined cost of duty and VAT.

Fines for shopkeepers who sell alcohol to under-age drinkers will be doubled to £20,000.

The plans, included in the Police Reform and Social Responsibility Bill, also include greater powers for police to close pubs and bars that attract trouble.

Almost half of all violent crime victims report that their attacker was under the influence of alcohol

A Home Office spokesman said they were determined to 'tackle the drink-fuelled violence and disorder which is blighting many of our communities'.

He added: 'We want to give police and local authorities more powers to strip problem premises of licences.

'A review of alcohol taxation and pricing will also be undertaken to ensure it tackles binge drinking without penalising responsible drinkers, pubs and important local industries.'

20 June 2010

THE TELEGRAPH

Chapter 2

PROBLEM DRINKING

Binge drinking

Information from TheFamilyGP.com

What is binge drinking?

It takes a lot less than you think

In general, binge drinking is the term used to describe drinking large amounts of alcohol in a short period of time, but specifically it means drinking more than twice the recommended daily alcohol limits.

The UK has one of the highest binge-drinking rates in Europe. Although it's often associated with inexperienced teenagers, approximately a quarter of adults in England admit to binge drinking.

During 2007-08, over 30,000 people in England were admitted to hospital with alcohol poisoning – 500 a week – mostly as a result of binge drinking. Of these, more than half were women, showing there is no gender gap when it comes to alcohol.

THEFAMILYGP.COM

To minimise the health risks associated with drinking alcohol, it is currently recommended that:

- men should drink no more than three to four units of alcohol per day.
- women should drink no more than two to three units of alcohol per day.

As an ordinary-strength pint of lager or bitter contains two units of alcohol, and a 175 ml glass of wine (14%) has two units, the units can quickly add up to levels which can cause harm to your health.

So, during the festive season, a woman drinking two glasses of wine at lunch, followed by a work party where she may drink a further two glasses, means that she has more than doubled what is regarded as safe drinking levels.

Continuing to drink puts her in danger of the short-term effects of binge drinking, as well as risking her future health.

How binge drinking affects your health

It can cause alcohol poisoning and more

Not surprisingly, binge drinking can cause serious health problems. These are caused by the high concentration of alcohol in the blood. Unlike food, which can take hours to digest, alcohol is rapidly absorbed by the body into the bloodstream.

Blood alcohol concentration

Your liver can only metabolise or process one unit of alcohol per hour. After a binge drinking session, there is nothing you can do to stop your blood alcohol concentration rising, as the alcohol in your stomach continues to be absorbed into the bloodstream.

This is why binge drinking is so dangerous. Your blood alcohol concentration can rise to levels which can kill – alcohol poisoning – even when you have stopped drinking, have fallen asleep or have passed out.

How binge drinking affects your body

Alcohol health problems are dose dependent, meaning they are directly related to how many units of alcohol you drink. Binge drinkers risk the same health problems associated with heavy drinkers, but even one night of binge drinking can cause:

- Double/blurred vision.
- Muscle pain and weakness.

- Stupor.
- Low body temperature (hypothermia).
- Low blood sugar levels (hypoglycaemia).
- Convulsions.
- Alcohol poisoning.
- Memory problems.

Repeated binge drinking also puts you at an increased risk of acute pancreatitis, gastritis, damage to the oesophagus, which may include acute haemorrhaging and high blood pressure.

Alcohol is a well-known depressant. It depresses the actions of the nerves which control the body's involuntary actions, including breathing, heartbeat and the gag reflex, designed to stop us from choking.

However, as alcohol is also a stomach irritant, drinking to excess may cause vomiting. If the gag reflex is depressed as a result of drinking too much, this can cause you to choke or inhale your own vomit, which could lead to death.

How binge drinking kills

Ways to handle alcohol poisoning

Binge drinking can kill overnight as a result of acute alcohol poisoning. Around 500 people a week in England end up in A&E hospital departments as a result of alcohol poisoning, and 157 died from it in 2007.

Warning signs of alcohol poisoning

The signs that someone has progressed from being obnoxiously to dangerously drunk include:

- Confusion.
- Can't be woken from sleep.
- Cold, clammy or blue skin.
- Vomiting.
- Convulsions.
- Slow breathing and/or irregular breathing (less than eight breaths a minute).
- Low body temperature (hypothermia).

If someone shows signs of alcohol poisoning, it's important they are not left alone and are looked after.

DO:

- Try to keep them awake.
- Try to keep them in a sitting or standing position if vomiting.
- Put them into the recovery position with their head to the side if they insist on lying down.
- Keep them warm to prevent hypothermia.
- Give them water if they can drink it.
- Stay with them. Even if they were initially responsive, the alcohol levels in their blood may be so high they will continue to get drunk whilst asleep, and may become unconscious.

DON'T:

- Give them coffee to sober them up. It doesn't work and will cause more dehydration.
- Leave them to sleep it off. They could inhale their own vomit and die.
- Allow them to lie on their back. They may vomit. Put them on their side, and use pillows or cushions to stop them rolling back.
- Walk them up and down to try to sober them up.
- Put them under a cold shower.
- Allow them to drink any more alcohol.

Don't wait for all of the symptoms to appear. If in any doubt, call 999 for an ambulance.

9 December 2009

- The above information is reprinted with kind permission from TheFamilyGP. com

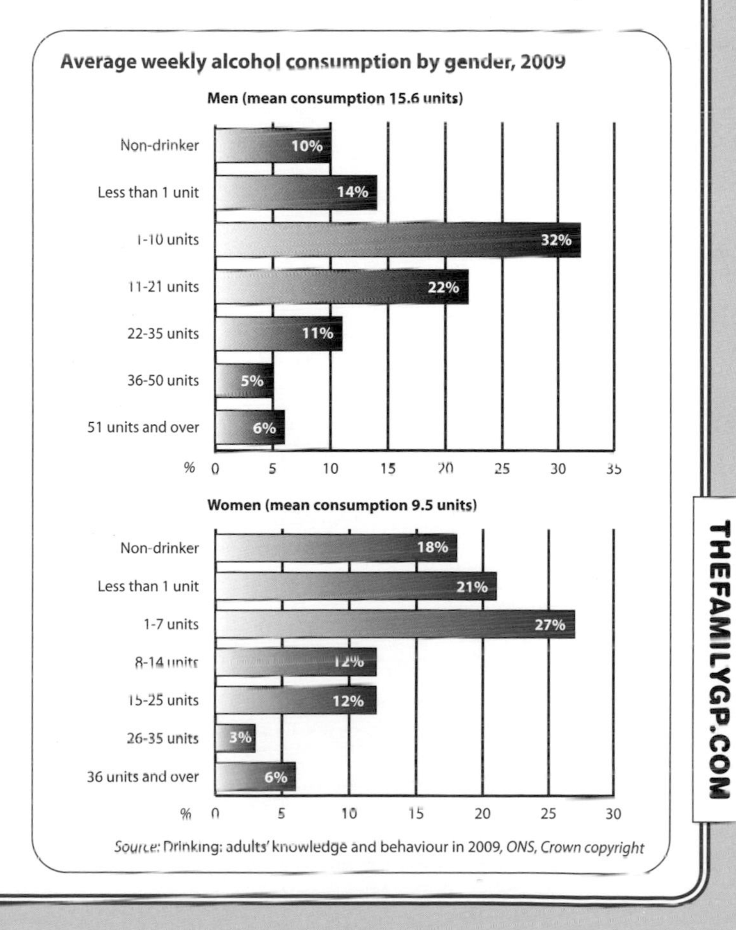

Affluent teenagers drink more, study shows

Many schools in better-off areas have a drinking culture, say researchers, while very disadvantaged students are least likely to have tried alcohol.

By Rachel Williams

Teenagers in better-off areas are more likely to consume alcohol thanks to a 'drinking culture' that has developed in some schools, according to a study published today.

Researchers found higher numbers of pupils were drinking at schools where lower proportions of students were eligible for free school meals and from ethnic minorities.

In a report on which young people are most likely to drink and what effect it has on their behaviour, a team from the National Centre for Social Research also discovered that children who are bullied frequently turn to alcohol.

Those who had been bullied in the last year were up to five times more likely to be drinking on most days than those who had not, a link not highlighted in previous studies and needing further investigation, the researchers said.

Overall, 56% of young people had had an alcoholic drink at the age of 14, and by 17 the figure was 86%, according to statistics from the centre's Longitudinal Study of Young People in England (LSYPE), which has been charting the progress of a 15,500-strong group of young people since 2004.

Young white people were the most likely to have tried drink, followed by mixed-race teenagers and those from black Caribbean backgrounds. Young people of Pakistani and Bangladeshi origin were the least likely to have done so.

Teenagers with parents who were unemployed, and those whose mothers had no qualifications from the UK, were less likely to have tried drinking than their middle-class counterparts.

'This seems to indicate that young people of very low social position may be less likely to try alcohol, possibly because it is less likely to be available in the home,' the researchers said.

They added: 'We found that young people who attended schools with a larger proportion of white pupils were more likely to have tried alcohol regardless of their own ethnic group, as were those who attended schools with fewer pupils who received free school meals (FSM), again regardless of their own FSM status.

'These results may indicate the presence of aspects of a "drinking culture" in some schools, whereby having a higher proportion of individual pupils who drink makes it more likely that those pupils who have characteristics that make them less likely to drink (for example, being from minority ethnic groups) are also more likely to try alcohol.'

Girls were more likely to try alcohol than boys. The researchers also found, apparently to their surprise, that young people who played sport or played a musical instrument were slightly more likely to drink than those who did not.

'It therefore appears that taking part in activities that might be considered to be self-developmental does not seem to deter young people from drinking,' they said.

A spokesperson at the Department for Education said: 'This research proves that alcohol can have a devastating impact on the lives of children and young people and causes much more than just health problems. The number of young people drinking alcohol is falling but it is important that the Government looks at what can be done to reduce this further and faster.'

24 June 2010

Thousands of children drinking seven pints a week

Thousands of children are drinking more alcohol each week than the recommended safe limits for adults, new figures have revealed.

By Rebecca Smith

A third of children aged 11 to 15 who drink, consume more than 15 units in a week, the equivalent of seven pints of lager or one and a half bottles of average strength wine.

It means 178,560 children in England are consuming more alcohol in a week than the recommended limit for an adult woman.

The news comes after the death of Gary Reinbach, 22, who was given only weeks to live after developing cirrhosis due to his heavy drinking from a young age. He had been denied a liver transplant.

Experts said more younger people are being struck with liver cirrhosis than ever before.

Just under one fifth of children aged 11 to 15 years old have drunk alcohol in the last week, or 558,000 children.

Although the proportion of children saying they have ever drunk alcohol is dropping, those who do drink are consuming more on average, the data has revealed.

The figures show the amount of alcohol children are drinking has on average increased from 5.3 units in 1990 to 9.2 units in 2007 when the calculations were changed.

Girls drank on average one unit a week fewer than boys throughout the period, the report entitled *Smoking, drinking and drug use among young people in England in 2008*, said.

In 2007 the number of units in a drink were recalculated to take into account the use of bigger wine glasses and stronger alcoholic drinks.

One pint of strong lager is now over 2.5 units and a large (250 ml) glass of average strength white wine is over three units.

Children aged between 11 and 15 years old drank on average 12.7 units in 2007 increasing to 14.6 in 2008, the report from the NHS Information Centre has found.

The recommended weekly limits for adult women is 14 units and 21 units for men.

Older children who drink were more likely to drink more, with more than one-third of 15-year-olds reporting that they drank more than 15 units in a week.

The annual survey took place across 264 English secondary schools and surveyed nearly 7,800 pupils aged 11 to 15, representing an estimated population of around 3.1 million pupils.

Just over half of children in the survey said they had ever drunk alcohol, a drop from a peak of more than 60 per cent in 2003. There was little difference between the sexes.

For the first time, the survey asked pupils about their household's drinking habits and found that those who live with people who drink are three times more likely to drink themselves.

The percentage of 11- to 15-year-old pupils who drank alcohol in the week prior to the survey was five per cent in non-drinking households, increasing to 31 per cent in households with three or more other people who drank alcohol.

More than eight out of ten children between 11 and 15 years old live with at least one drinker.

Smoking was also more common in children who live with smokers. Just three per cent of children whose parents did not smoke said they themselves smoked regularly compared with more than a fifth of children who lived with three or more smokers.

Around a half of children who drink said they drink at home.

Two-thirds of 11-year-olds who drink said they drank with their parents.

Professor Ian Gilmore, President of the Royal College of Physicians said: 'These figures show very clearly that we need to tackle teenage drinking from every possible angle.

'While It's absolutely right to focus on issues of price, and illegal sales it is equally vital that parents, teachers and other role models acknowledge that their own behaviour shapes children's attitudes towards alcohol.

'In a sense this cuts to the heart of what we mean when we talk about our drinking culture. If we are to

THE TELEGRAPH

encourage healthier attitudes among the young, adults need to think more carefully about the examples we set for our children.'

Alcohol Concern Chief Executive Don Shenker said: 'Today's figures are very worrying. We're seeing a slight decline in the number of children who drink, but those who do drink are drinking much more.

'Too many young people are now drinking at or above safe adult levels, yet their bodies are less able to cope with the harm alcohol can cause.

'We've already seen an almost one thousand per cent increase in liver cirrhosis deaths in the 25-44 age group. This is impacting our health services and the lives of families across the UK.

'Time and time again we hear from families that alcohol is too cheap and too easy for young people to get hold of. If the Government really wants to tackle alcohol misuse in this country, then it needs to get tough and follow the Chief Medical Officer's recommendations for a minimum price for alcohol.'

Alison Rogers, Chief Executive of the British Liver Trust said: 'The influence that parents have on their children's drinking is incredible and something that shouldn't be overlooked – this report clearly shows this. Sadly, we know that increasing numbers of young people are suffering serious health problems, including fatal liver damage, due to drinking too much alcohol.'

Children's Minister Dawn Primarolo said: 'I am pleased that today's statistics show a continued decline in the number of young people drinking alcohol; however, it is disappointing to see that those who choose to drink, are drinking more than last year.

'But what today's findings also demonstrate is the need to make sure guidance and information is suitably targeted at parents as well as their children. Ultimately we want to see young people waiting longer before consuming alcohol, but when they do decide to drink, we need to make sure that they and their family have clear support and advice from professionals.'

23 July 2009

Fewer kids drinking – but don't hold the front page

By Leo Henghes

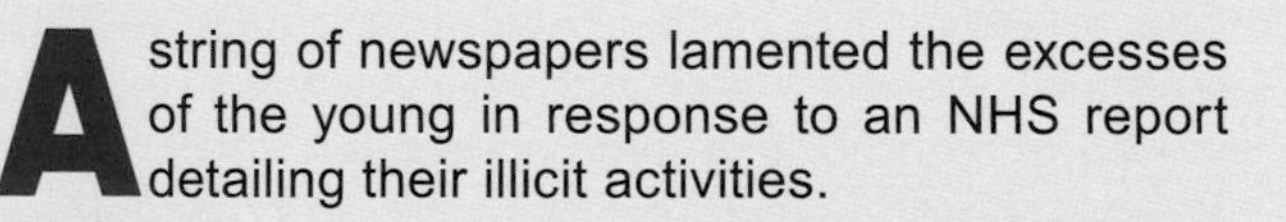

A string of newspapers lamented the excesses of the young in response to an NHS report detailing their illicit activities.

Smoking, drinking and drug use among young people in England in 2008, from the Information Centre for Health and Social Care, provides an insight into the habits of those aged between 11 and 15. The overall message was reasonably positive – fewer youngsters are drinking even if the minority who do are downing larger quantities – and that might explain why some newspapers were selective in their choice of statistics and why others chose not to run the story.

Some were particularly keen to highlight the fact that 550,000 youngsters said they had drunk in the last week, but failed to contextualise the figure appropriately. Indeed, the proportion of children who said they had drunk alcohol in the last week has fallen to 18 per cent, from a peak of 26 per cent in 2001 and 20 per cent in 2007.

The *Daily Mail* failed to clarify that an average consumption of 14.6 units quoted in the report applied only to those 18 per cent of children who admitted to drinking in the last week, not an average of all children. It added that 'the amount they knocked back...was twice as high as in 1990.' True, the report does show that, among those who drink, the average consumption is increasing. But it also specifically states that figures from this year should not be compared with previous years, because of changes in the way units drunk by children are measured to take into account bigger wine glasses and stronger drinks.

The Sun and the *Daily Express* focused on children's drug use, linking the report with other newly released figures from the Department of Health that 60 children were admitted to hospital with acute cocaine poisoning in 2007–08.

But the NHS report contradicted the *Express* headline claiming a large rise in drug use. What it actually showed was that while 1.7 per cent of children aged 11-15 had taken cocaine in the last year, that was a reduction on the 1.8 per cent recorded the year before. As for *The Sun*, its assertion that it was 'staggering' that eight per cent of children admitted to taking drugs within the last month showed a lack of perspective, as it was actually the lowest percentage reported this decade, with a fall from ten per cent the previous year.

And the reduction in smoking amongst children was ignored by almost all the papers, despite this positive development taking up a third of the actual report.

6 August 2009

⇨ The above information is reprinted with kind permission from Straight Statistics. Visit www.straightstatistics.org for more information.

Pain and anger are the hidden burden for children with an alcoholic parent

At the start of Children of Alcoholics Week, victims talk about their shame, loneliness and guilt with nowhere to turn for help.

By Tracy McVeigh

When Mary Smith pulled the car off the road to answer her mobile and hear the news of her father's death, she felt just 'a calm relief'.

'Really, I had lost my dad many years before. His mind had gone at least four years before,' said Smith (not her real name). 'Sometimes I think about what we went through and I can't quite believe that we got through it. There were a lot of bad times.'

Her father was an alcoholic who drank himself to death. All the help his young daughters and his wife tried to get him, from detox programmes, to rehab, to psychiatric sessions, had failed. 'He chose to drink, and he chose that over us. It took me a long, long time to accept he had a disease, but my anger had gone before he died.' Smith, now 27, spent her late teenage years trying to protect her younger sisters, support her mother and get help for her father. There was little time for her to enjoy her youth.

'I did take the brunt of it. No one should have to beg their mother to get a divorce, or their father to stop drinking and choose them over alcohol.'

Smith and her three sisters are among the 3.6 million people in Britain who have had their childhoods scarred by the drinking of one or both of their parents. The latest research suggests that almost one million children in Britain are today living with an alcoholic parent, but campaigners say the true number is far higher, obscured behind the front doors of a society where drinking is so much the norm that even social workers don't take it seriously.

'If a social worker goes to a home where a parent is smoking heroin or cannabis, action is taken straight away,' said Don Shenker, chief executive of Alcohol Concern. 'But if alcohol is being taken, they won't even bother recording it. We are finding a lot of problems with alcohol abuse not being included in case notes.'

He said that far too much policy consideration was given to drugs users, while the far bigger picture, the problem of families living with alcohol addiction, was being overlooked.

'It's the forgotten issue, but it's actually on a far greater scale than drug addiction,' said Shenker, who is calling for better training for social workers to recognise alcohol-related issues. 'The support services are not there, certainly not working with the families. Even where alcohol dependency is being treated, no one is saying to these people, "And do you have children? And how is your drinking impacting on them?"

'It's just not talked about... The kids don't want to talk about it because they will feel guilty, the parents who are drinking [feel they] have something to hide, and society doesn't really want to come out against alcohol. So there's this mismatch where families at risk through drugs get strategies, but those at risk through alcoholism do not.'

But there is now a growing support community for children living with alcoholic parents, set up and run mostly by people who have been through the experience themselves.

This week marks the second Children of Alcoholics Week, run by the National Association for Children of Alcoholics (Nacoa), to raise awareness of the difficulties faced by millions who are, or were, parented by alcoholics. The

THE GUARDIAN

charity saw the number of calls to its helpline double to 38,000 during a 12-month period over 2007 and 2008, and believes the hidden issue is blighting many more young lives than previously thought.

The campaign is backed by celebrities including Lauren Booth, Calum Best, Virginia Ironside and Geraldine James, all themselves children of alcoholics.

'As children, we were never allowed to talk openly about our mother's drinking,' said actress James, who said she desperately wished there had been help. 'I remember my brother being slapped very hard when he asked my father if she was drunk – "Never use that word in this house again."

'I grew up feeling ashamed, frightened, lost, guilty and lonely; feeling unconfident, unsafe, unlistened to, unprotected, unloved, unlovable; feeling there was no one there, inside or out. But there was literally no one to turn to.'

It is a sentiment shared by Emma Spiegler, 26, a Nacoa trustee who started Children of Addicted Parents in 2006 after realising that children living in shame and guilt would find it easier to use online forums where they could have anonymity. Her own family was 'pulled apart' by her mother's drinking.

'There was nothing out there at all, and I know how lonely and isolating it can be. It's all about loyalty and the guilt and the shame, it's just so massively hidden.

'Mum was a social drinker from when I was about four upwards. She was in rehab when I was ten. But she drank at night, so it wasn't until I was about 13 and starting to stay up later that I realised, and realised it wasn't normal, the wine bottles lying around and her falling asleep downstairs. Unlike a lot of people, my mum wasn't a violent drunk or aggressive, just emotional, so she would cry and get very vulnerable and say, "I've been a bad mother and I've messed everything up", and I'd have to comfort and emotionally reassure her.

'I was a very angry child, really unhappy, and she was too involved in her relationship with the bottle to ever be there. I didn't tell anyone and I don't know why. I saw my dad at weekends, my parents were divorced, but I never talked to him about it. I think children just don't. I lived my life for my mum.'

Spiegler's mother has now been sober for four years: 'I don't think she really realises what happened.'

Statistics back the damage that drinking does inside Britain's families. Growing up in an alcoholic household was inextricably linked to abuse, with 55% of domestic violence incidents happening in alcoholic homes and drink being a factor in 90% of child abuse cases. The NSPCC reports that one in four cases of neglect reported to them involves a parent who drinks.

A study by the Priory Clinic group in 2006 found that children who grow up with alcoholic parents bear emotional, behavioural and mental scars and their early lives were characterised by chaos, trauma, confusion and shame and, quite often, sexual and physical abuse. Studies have also shown that a third of daughters of alcoholics experienced physical abuse and a fifth sexual abuse – up to four times higher than in non-alcoholic homes.

The report said that children reacted in one of three ways: they became withdrawn, went into denial or used the experience to benefit themselves by becoming stronger. They also struggled to develop strong personal relationships. The report added: 'Their feelings about themselves are the opposite of the serene image they present – they generally feel insecure, inadequate, dull, unsuccessful, vulnerable and anxious.'

Dr Michael Bristow, an addictions expert at the Priory, said: 'There is a widespread misconception that addiction is all about the addict, that it is solely the addict who suffers. The reality? Alcoholism affects the adult alcoholic's entire family, particularly the children.'

Partly because of their genes, children of alcoholics are four times more likely to become alcoholics than the one in 20 average in the population that currently have the condition, and 50% will end up marrying an alcoholic.

What the late Mo Mowlam, who was patron of Nacoa and whose father was an alcoholic, called the 'hidden suffering' of UK families is getting worse as excessive drinking becomes more of an issue.

'If this could happen to me and my family, how many other families must be suffering in silence?' said Mary.

Nacoa helpline – 0800 358 3456

Alcohol's hidden costs

- More than two million adults in the UK claim to have been raised by parents who drank too much.
- 33% of children of alcoholics go on to develop related problems in adulthood.
- Out of 1,000 adults, 47 are likely to be dependent on alcohol – double the amount of people who are dependent on illegal drugs.
- A chemical known as THIQ only occurs in the brain of an alcoholic. It has been found to be more addictive than morphine and an effective painkiller.
- Some 40% of violent crime, 78% of assaults and 88% of criminal damage cases are committed while the offender is under the influence of alcohol.

⇨ This article is an extract from a piece which appeared in the *Observer* on 14 February 2010.

Crime and disorder

Alcohol-related crime costs the country £7.9bn per year.

About half of all cases of violent crime are thought to be alcohol-related – there is a clear link between alcohol misuse and crime but drinking does not cause crime.

Increasingly, many people feel uncomfortable in town centres as they fear the disorder associated with large numbers of people drinking to excess. A 2009 YouGov survey conducted for Alcohol Concern showed that this could be as many as a third of the population who thought that town centres had become 'no-go' zones at night due to alcohol-related problems.

About a fifth of alcohol-related crime is committed in or around licensed premises and there is a link between the density of licensed premises and crime

About a fifth of alcohol-related crime is committed in or around licensed premises and there is a link between the density of licensed premises and crime. There is a strong need to promote sensible retail practice in the sale of alcohol and sufficient policing of the night-time economy. Encouraging safer drinking and making sure that those who commit alcohol-related crime benefit from addressing their drinking rather than just being punished, would go some way to tackling the problems we face.

So far the voluntary retail regulations operated by clubs and pubs have failed to address these issues and Alcohol Concern has called for a Mandatory Code on Alcohol Sales to make practice safe and consistent which will ensure town centres are more pleasant environments for everyone to enjoy.

Alcohol Concern believes that a public health objective should be added to the Licensing Act, as it has in Scotland, enabling Licensing Authorities to make decisions about licenses based on the overall health and well-being of an area. Local authorities need to be given greater freedom than the current Licensing Act (2003) allows; they need to be able to make decisions, in terms of changing, refusing or withdrawing licenses on the basis of what is the best interest and health of local communities.

Arrest referral schemes should be properly funded and rolled out across the country. These have been shown to be effective and would effectively reduce alcohol-related crime in the long term.

Tackling the cheap price of alcohol both in 'on' and 'off' premises would have a significant impact on the level of alcohol-related crime and disorder. For example, research shows that a 50p minimum per unit price would result in 10,000 fewer violent crimes.

⇨ The above information is reprinted with kind permission from Alcohol Concern. Visit www.alcoholconcern.org.uk for more information.

THE JOY OF A SOCIAL DRINK...

ALCOHOL CONCERN

Priory research highlights alcohol problem

Information from the Priory Group.

Government guidelines for safe drinking are having little effect on the nation's drinking culture, according to an ICM poll carried out on behalf of the Priory Group.

When questioned one in three women and one in five men did not know the number of units specified in the national guidelines. Almost half of over 55s were not aware of the much publicised safe alcohol consumption limits.

The research also found that young people are not the worst offenders for excessive drinking, with those aged 35-44 the most likely to drink too much

The heaviest drinkers are found in 55- to 64-year-olds where six per cent habitually drink more than 43 units a week.

Regionally, the North of England has the highest number of 'over the limit' drinkers, closely followed by Scotland. Wales and the South West of England fare best with only eight per cent.

Highlights of the research findings:

- 21% of men regularly drink in excess of the national guidelines.
- 15% of women regularly drink in excess of the national guidelines.
- 21% of 35- to 44-year-olds regularly drink in excess of the national guidelines.
- 12% of 25- to 34-year-olds regularly drink in excess of the national guidelines.
- 11% of 45- to 54-year-olds regularly drink in excess of the national guidelines.
- 20% of men did not know the correct number of units in the guidelines.
- 29% of women did not know the correct number of units in the guidelines.
- 49% of over 55s did not know the correct number of units in the guidelines.

Regional findings for the number of people drinking in excess of the national guidelines:

- **North of England** 17%
- **Scotland** 16%
- **Midlands** 14%
- **South East** 11%
- **Wales and the South West** 8%

More than half the people questioned preferred to drink at home or at a friend's house. However, 78% of 18- to 24-year-olds favoured a pub, club or bar.

Experts at the Priory Group have put together a number of simple questions to help individuals ascertain if alcohol is starting to become a problem.

1. Are you worried you're drinking too much?
2. Have friends or family expressed concerns about your drinking habits?
3. Has drinking affected your work, family or personal relationships?
4. Can you drink a lot without becoming drunk?
5. Do you experience blanks in your memory when drinking?
6. Have you ever tried to stop drinking, but returned to it after just a few days?
7. Do you feel shaky, sweaty or anxious a few hours after your last drink?
8. When drinking, do you find yourself doing things you normally wouldn't do?

Anyone answering yes to several of the questions may be at risk.

Dr Mark Collins, consultant at the Priory Group said: 'What these figures clearly show is that large numbers of people are still drinking more than is good for them.

'Despite high-profile advertising campaigns the national guidelines are still a mystery to many people and this is worrying. When people were asked how many units they thought they drank this was universally a much smaller number of units than the actual amount consumed.

'The recommended limits are not designed to spoil people's fun but to ensure that they do not drink amounts that can be physically and mentally damaging.

'Many people enjoy a drink without it negatively effecting their day-to-day lives, but with drinking a part of many people's social or working environments, it can be easy to lose track.

'Anyone worried about their drinking habits should seek advice and support from their GP in the first instance. If necessary, GPs can refer you to centres like ours for further assessment and help.'

27 October 2009

- Information from the Priory Group. Visit www.priorygroup.com for more information.

PRIORY GROUP

Drink driving

It is an offence to drive, attempt to drive, or be in charge of a motor vehicle on a road or public place if the level of alcohol in your breath, blood or urine exceeds the prescribed limit.

The law specifically states that your alcohol level at the time of any alleged offence is presumed to be the same, or not less than, the result of the analysis of the breath, blood or urine sample. It may be possible to challenge this if you can show that:

- ⇨ you had consumed alcohol after you stopped driving but before the specimen for analysis was provided; and
- ⇨ had you not taken that alcohol you would not have been over the limit; and
- ⇨ your ability to drive would not have been impaired.

Note: If the offence alleged is being 'in charge', it is a defence if you can show that there was no likelihood of you driving while you were over the limit.

Legal limit

The legal limit of alcohol in the body is:

- ⇨ 35 micrograms (μg) per 100 millilitres of breath.
- ⇨ 80 milligrams (mg) per 100 millilitres of blood.
- ⇨ 107 milligrams per 100 millilitres of urine.

Failing to provide a sample

An offence is committed if you:

- ⇨ fail to provide a specimen of breath for a preliminary breath test; or
- ⇨ fail to provide a specimen of br[illegible] the police station when requested [illegible] reasonable excuse*); or
- ⇨ refuse to provide a specimen.

* Generally, for an excuse to be considered it must relate to a physical or mental inability to provide the sample. In law:

- ⇨ imposing conditions (e.g. delaying a sample being given until a solicitor arrives) amounts to a refusal.
- ⇨ the police are under no obligation to delay the process of taking samples and unless there is a relevant medical reason, our advice is that you should never refuse to give a specimen, regardless of whether you feel your refusal would be justified or not.

If you refuse and subsequently a court does not accept your reasons, it will be too late to avoid the offence of refusal. A conviction for failing or refusing to give a sample at the police station would make you liable to an automatic disqualification.

Being unfit to drive

It is an offence to drive, attempt to drive, or be in charge of a mechanically propelled vehicle on a road or public place while unfit through drink or drugs.

You would be considered unfit if your ability to drive properly is impaired (even if the amount of alcohol in your body is within the prescribed limit). Evidence of being unfit may be found, for example, in erratic driving, the occurrence of an accident, or your condition.

⇨ Information from the Automobile Association. Visit www.theaa.co.uk for more information.

Half of children see parents drunk

Half of children (50%) think they have seen their parents drunk at some time, according to a survey just published.

Almost a third (30%) of children feel scared when they see adults drinking, the study for CBBC's *Newsround* discovered.

Nearly three-quarters of the children questioned (72%) said their parents drink alcohol.

Of those, seven out of ten (70%) thought they had seen them drunk.

Just less than half of the children (46%) surveyed thought that adults should not drink in front of children, while 32% thought this was all right, and 22% were not sure.

When given a list of words to describe how they felt when they saw adults drinking, almost half the children (47%) said they were not bothered. But 30% said it made them feel scared.

Eight in ten children (81%) surveyed who had seen adults drinking said they noticed one or more changes in the way they behaved.

Of those, almost a quarter (24%) said it made adults act stupid or silly; a fifth (20%) said they became angry and aggressive; another fifth (20%) said they became happy and funny; 19% found them to act strangely or in a different way; 18% said they became loud and swore; and 17% said they became dizzy or fell over.

The results were revealed on CBBC's *Newsround* on BBC1 as part of a wider exploration of the issue of alcohol and how adults' drinking affects children's lives.

Damian Kavanagh, Controller of CBBC, said: '*Newsround* has a great track record in covering complex issues in an engaging and helpful way. There has been much discussion about levels of drinking but the social impact is rarely explored from a child's point of view.'

5 July 2010

Cutting drink-drive limit could save lives, NICE study finds

Information from National Institute for Health and Clinical Excellence.

Cutting the drink-drive limit for motorists in England and Wales by almost half could prevent around 3,000 road injuries and 145 deaths in the first year alone, a study by NICE has found.

Currently, the legal alcohol blood concentration limit for drivers is 80 mg of alcohol per 100 ml of blood. But reducing the limit to 50 mg/100 ml, in line with most countries in Europe, will result in major improvements to road safety.

When a limit of 50 mg/100 ml limit was introduced in 15 countries in Europe, it collectively led to 11.5 per cent fewer alcohol-related driving deaths among 18- to 25-year-olds – the age group most at risk of being involved in accidents.

Over 400 people were killed in drink-driving accidents, accounting for 17 per cent of all road fatalities, according to UK figures from 2008.

The British Medical Association (BMA) has already called for a reduction in the drink drive limit because there is clear evidence that a driver's reaction times and motoring skills deteriorate after even a small amount of alcohol – and gets worse with increased alcohol consumption, and that the risk of being involved in a collision rises significantly once the blood alcohol level rises above 50 mg/100 ml.

The Royal College of Physicians and the Alcohol Health Alliance UK have also voiced concern that the blood alcohol limit should be reduced to 50 mg, arguing that the limit has been accepted by all EU countries with the exception of the UK and Ireland, and Ireland is currently in the process of reducing the level to 50 mg/100 ml.

The findings of this latest study, carried out by NICE on behalf of the Department for Transport, will be considered as part of the Department's review of drink and drug driving law led by Sir Peter North.

For the study, NICE conducted a review of the best available international evidence from the USA, Australia and several European countries, and looked at drink-driving patterns and the associated risk of being injured or killed in a road traffic accident.

Modelling methods developed by researchers at the University of Sheffield were used to estimate the potential impact of lowering the drink-drive limit in England and Wales.

'Overall, the international evidence indicates that lowering the BAC limit from 80 mg to 50 mg could reduce the number of alcohol-related deaths and injuries in the UK,' said Professor Mike Kelly, Director of Public Health at NICE.

'Not only could it have a positive impact on those who regularly drink well above the current limit before driving, but it also has the potential to make everyone think twice about having a drink before they decide to drive somewhere.'

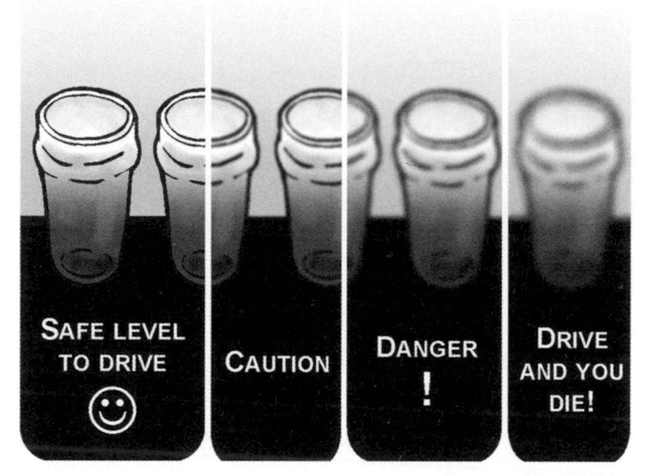

However, for this measure to effectively change people's attitudes to drink-driving and improve road safety in the long term, it must be supported by ongoing publicity, as well as visible and rapid enforcement. According to the evidence, current efforts do not appear to be a strong enough deterrent as many drivers do not believe that they will ever be caught or sanctioned, warned Professor Kelly.

'We therefore hope that our findings will provide the Government with an important basis for informing its policy considerations on changing the current drink-driving legislation,' he added.

16 June 2010

⇨ National Institute for Health and Clinical Excellence (2010). *Cutting drink-drive limit could save lives, NICE study finds*. London: NICE. Available at www.nice.org.uk. Reproduced with permission.

GPs to quiz NHS patients over alcohol int

Watchdog pleads for minimum pricing.

All NHS patients face being quizzed about their drinking habits by their GP under tough new alcohol guidelines recommended by the NHS's watchdog.

Patients could be asked how much and how often they drink when they register with a GP, attend Accident and Emergency or receive advice about their medication or sexual health.

The guidance from the National Institute for Health and Clinical Excellence – the Government's health advisory body – also recommends setting a minimum price for alcohol to combat Britain's binge-drinking epidemic.

In addition, a crackdown on cheap drink, booze cruises and a possible advertising ban are included among measures put forward to make alcohol more difficult to buy.

Professor Eileen Kaner of Newcastle University chaired the NICE group of experts who made the recommendations.

She said: 'It should be a common medical practice to ask about alcohol where alcohol could be a contributory factor or (where patients have) a condition where alcohol is likely to be a factor, such as sleep disturbance and hypertension.'

How the cost would rise

A minimum price of 50p per unit of alcohol would mean:

A bottle of wine (12% alcohol)	£4.50
500 ml can of lager (4%)	£1.14
Six cans of lager	£6
Ten cans of lager	£10
Two-litre bottle of cider (7.5%)	£7.50
70 cl bottle of vodka or rum (37.5%)	£13.13
A pint of stout (4.1%)	£1.17
A pint of ale (4%)	£1.14
A litre of rum (40%)	£20
A 275 ml bottle of 'alcopop' (5%)	£0.68

Benefits, say NICE, include 3,393 fewer deaths each year, 97,900 fewer NHS admissions and 45,800 fewer crimes.

Those patients thought to be drinking too much could be advised to stay off alcohol for one or two nights a week, have one or two fewer drinks or eat before they go out drinking, she added.

The guidance is the strongest call yet for a radical shake-up in the nation's 'unhealthy relationship' with alcohol.

It will severely test the coalition Government's approach to excess drinking.

The Tories have already pledged to stop supermarkets and off-licences selling alcohol below cost price to encourage trade but have stopped short of backing a minimum price, fearing an industry backlash.

However, the Lib Dems have a minimum price policy in their manifesto. NICE stopped short of saying what this price should be, although the British Medical Association is backing a 50p per unit minimum.

Research shows increasing prices has the biggest effect on the heaviest consumers and on young people, who spend a relatively high proportion of their income on alcohol.

The guidance from NICE also says there should be cuts in the amount of drink holidaymakers and 'booze cruisers' are allowed to import.

The number of outlets selling alcohol could be reduced, along with opening hours, while councils should restrict new licences in 'saturated' areas to cut crime and disorder.

Shops selling to those who are under age or clearly drunk should face penalties or closure, says the guidance, while a complete ban on advertising would protect young people.

Professor Mike Kelly, NICE's public health director said the annual toll of excessive drinking costs the NHS £2 billion, leads to 500,000 related crimes, 17 million lost working days, 1.2 million violent incidents and just under 15,000 alcohol-related deaths.

However, Tory MP Peter Bone, a former member of the House of Commons Health Select Committee, said: 'My main problem with minimum pricing is that it increases the profits of supermarkets and does not increase Government revenue.

'My other problem is that you will not combat yobbish behaviour in this country just by putting up the price.

'It is not going to solve the problem. We need a cultural change in this country.'

2 June 2010

⇨ This article first appeared in the *Daily Mail*, 2 June 2010.

THE DAILY MAIL

Knowledge of drinking guidelines does not equal sensible drinking

***Information from the* Lancet.**

By Antony C Moss, Kyle R Dyer and Ian P Albery

The report *Reducing alcohol harm: health services in England for alcohol misuse* concluded that the UK's current approach to alcohol-related policy, prevention and treatment is ineffective and uncoordinated. Among the recommended platforms for action were revision of policies for the pricing and promotion of alcohol, mandatory labelling and revised sensible drinking guidelines. The focus on clear labelling and guidance carries with it the assumption that providing such information and facilitating its comprehension will reduce irresponsible drinking behaviour. In our opinion, such efforts are not sufficient to affect drinking behaviour; to demonstrate this, we explored the relation between drinking behaviour and knowledge of sensible drinking guidelines in a sample of medical students at a London institution.

A total of 203 social drinking undergraduate medical students (66% female, mean age 25 years [range 18-52]) completed a short online questionnaire; 114 (56%) reported heavy and binge drinking (defined as a score on the Alcohol Use Disorders Identification Test [AUDIT] of =7 for men and =5 for women) in the past six months. Further AUDIT data showed that 158 (78%) consumed alcohol hazardously, 83 (41%) were at risk of alcohol dependence, 131 (65%) reported experiencing alcohol-related harms, and 49 (24%) reported at least one occasion where they were not able to stop drinking once started. More than half the respondents classified as heavy and binge drinkers (63 [55%]) were not considering engaging in sensible drinking. This high-risk drinking behaviour occurred despite the fact that 199 (98%) of the overall sample correctly reported current UK drinking guidelines, and all had received a structured curriculum incorporating specific education on alcohol and other drug misuse.

Policies that focus on factors associated with the purchase and consumption of alcohol, and, where appropriate, enforce legislative measures, seem effective and cost-effective in reducing alcohol-related harm. However, we argue that these population-wide interventions are not a panacea. Advances in our understanding of intoxication and the link between alcohol consumption and the more complex societal harms of drinking, such as antisocial behaviour and violent crime, suggest that such measures would be ineffective in moderating the behaviour of individuals once they have already consumed alcohol. There remains a need for policy-driven research for those who will continue to experience and cause alcohol-related harm despite population-wide interventions.

To achieve this, we call for a Government-led consultation throughout the broader scientific community, involving key stakeholders from Government and industry, to guide the development of a coordinated national evidence-based policy and associated research agendas.

114 (56%) medical student undergraduates surveyed reported heavy and binge drinking in the past six months

Note

We declare that we have no conflicts of interest. Funding for the collection of the data reported was provided by grant RHB0017 from St George's, University of London awarded to ACM. Ethics approval was provided by the Wandsworth Research Ethics Committee.

10 October 2009

⇨ The above information is reprinted with kind permission from the *Lancet*. Visit www.lancet.com for more information.

THE LANCET

The real cost of alcohol

In a recession, when is alcohol too cheap? Don Shenker asks how the Government should tackle the problems associated with cheap booze.

In the time it takes you to read this article around 15 people will have been admitted to hospital with an alcohol-related complaint. More than 810,000 people were admitted to hospital in 2007-8 with alcohol-related illnesses – that's around 1.5 per minute. The problem is that while nearly everyone agrees this figure is too high and a burden on the NHS, the Government is deeply divided as to how to resolve it.

The cause of these astonishingly high admission rates – and rising by around 80,000 a year – has been squarely laid at the door of the price of alcohol. Alcohol is now around 70 per cent more affordable than it was in 1980, but taking action to tackle cheap alcohol is not a vote winner – not in a recession and not with an election forthcoming. When the Chief Medical Officer suggested introducing a minimum price per unit of alcohol he was roundly criticised in the press. After all, surely the majority of drinkers act responsibly and shouldn't have to pay more for the 'sins of the minority'? This has been the mantra of the drinks industry and certainly of the main two political parties, nervous of appearing too 'nannyish' and not wanting to upset voters who enjoy their two-for-one wine deals at the local supermarket.

Fortunately, we now have more evidence than ever of the impact of cheap sales of alcohol on harm and of the potential long-term health, crime and economic benefits of raising prices. Two recent reports from the School of Health at the University of Sheffield on alcohol price, consumption and harm have cast a light on the real cost of cheap alcohol and opened up the debate on how to solve this by introducing a minimum price per unit of alcohol sold. A minimum price would end loss-leading by supermarkets and raise the price of the cheapest drinks.

Raising the price of alcohol to 40p per unit would halve the number of people admitted to hospital this year. In addition, there would be 16,000 fewer crimes committed and 100,000 extra days worked. These are staggering numbers. The Sheffield research shows, importantly, that it is young drinkers, binge drinkers and harmful drinkers who tend to choose cheaper drinks. Consumption levels in this group are the most affected by price increases and decreases, and, crucially, this group consumes a disproportionately large amount of alcohol.

In other words, it is young, binge and heavy drinkers who are the drinks industry's best customers. They would reduce their drinking the most if a minimum price on alcohol was introduced and the vast harms associated with their drinking would diminish. According to the Sheffield report, a ten per cent increase in price, for example, would cut consumption among 11 to 18-year-olds by 35 per cent. The same price increase would only amount to a ten per cent decrease in consumption among moderate drinkers.

This is because moderate drinkers tend not to be so choosy about the price of their drinks – they do take advantage of price offers, but similarly are just as likely to spend a bit more. Whereas heavy drinkers would end up paying on average £2.60 more per week for their drinks if a minimum price of 40p per unit was introduced, a moderate drinker would only pay 11p more. This level of price increase alone would save the country £7.8bn over ten years.

You might well assume that without heavy drinking the profits of the drinks industry would diminish significantly, as higher minimum prices force heavy drinking groups to spend much less on alcohol. Ironically, if a minimum price were introduced, the industry would continue to profit, as although the volume sold would decrease, prices would be higher. In fact a 40p minimum price would yield an extra £633m, split between the on and off trade.

Although higher alcohol taxes would be the ideal way to raise the cost of alcohol and allow public funds to be spent on alcohol services, the off trade will always absorb the increases unless they are forced not to sell below cost. The Police and Crime Bill, currently being debated in parliament, will establish a mandatory code of practice to tackle irresponsible sales from the on trade, such as 'ladies drink free' offers or £1 shooters. While this is welcome, dealing with loss-leading in the off trade, which allows supermarkets to charge less than £3 for a three-litre bottle of six per cent cider, is much trickier unless there is a minimum price per unit.

Alcohol Concern has long argued that alcohol needs to be made less affordable. The obvious solution to this is to raise taxes, which we have argued for. But to shift the trend of bulk-buying and deep discounting, something else is needed to force the industry's hand. Scotland is looking to introduce a minimum price on alcohol – will England follow as it did with smokefree legislation? If we want to reduce alcohol harm, then let's hope so.

18 May 2009

➪ Information from *Drink and Drugs News*. Visit www.drinkanddrugsnews.com for more information.

DRINK AND DRUGS NEWS

NICE recommends action to reduce alcohol-related harm

Alcohol needs to be less affordable and less easy to buy if we are to save thousands of lives each year, says new guidance from NICE.

Around one in four men and women are currently drinking dangerous amounts of alcohol that are causing, or have the potential to cause, physical and mental damage. To help create an environment that supports lower-risk drinking, the National Institute for Health and Clinical Excellence (NICE) has published guidance today (2 June) outlining the most effective measures that can be taken to lower the risks of alcohol-related harm.

As well as the detrimental health consequences, there are a number of other knock-on effects that alcohol misuse can have, such as on antisocial behaviour, crime, costs to the NHS, relationship breakdown and work absenteeism.

Professor Mike Kelly, Public Health Director at NICE said: 'Alcohol misuse is a major public health concern which kills thousands of people every year and causes a multitude of physical, behavioural and mental health problems. What's more, it costs the NHS over £2 billion annually to treat the chronic and acute affects of alcohol – this is money that could be spent elsewhere to treat conditions that are not so easily preventable.

Alcohol should be made less affordable by introducing a minimum price per unit. This price should be regularly reviewed so that alcohol does not become more affordable over time

'Based on the international evidence, it is clear that policy change is the best way to go about transforming the country's unhealthy relationship with alcohol and prevent people from getting to the stage where they are drinking worryingly large amounts. Our guidance looks at a number of ways that the Government can consider doing this, from reducing the affordability and availability of alcohol, to looking at how advertising affects children and young people.

'It is NICE's job to improve the health of the population, and there is no doubt that if these measures are taken forward, that they will significantly decrease alcohol consumption and thereby offset some of the serious social, economic and physical health problems that arise as a consequence of drinking too much.'

Among the policy recommendations, NICE advises that:

- Alcohol should be made less affordable by introducing a minimum price per unit. This price should be regularly reviewed so that alcohol does not become more affordable over time.
- Alcohol should be made less easy to buy; for example, by reducing how much individuals are allowed to import from abroad, or by reducing the number of outlets selling alcohol in a given area, or the days and hours that it can be purchased.

- Applications for new licenses to sell alcohol should be based on the number of outlets in a given area, the proposed business times and the potential impacts on crime, disorder and alcohol-related illnesses and deaths.
- 'Protection of the public's health' should be added to the current licensing objectives, as is already the case in Scotland. This would mean that premises would have a legal obligation to consider the health of their customers when selling them alcohol.
- The current advertising regulations should be strengthened to minimise children and young people's exposure to alcohol products. A complete ban on alcohol advertising should also be considered to protect these high-risk groups even more, as is the case with tobacco products.

Professor Anne Ludbrook, Guidance Developer and a Health Economist said: 'Alcohol is much more affordable now than it ever has been – and the price people pay does not reflect the cost of the health and social harms that arise. When it is sold at a very low price, people often buy and then consume more than they otherwise would have done. It is a dangerous pattern which many people have unknowingly fallen into.

The current advertising regulations should be strengthened to minimise children and young people's exposure to alcohol products. A complete ban on alcohol advertising should also be considered

'There is a strong body of evidence from around the world to show that making alcohol less affordable will reduce its consumption. This will in turn, improve the overall health of the population.

'NICE's recommendation to introduce a minimum price per alcohol unit is a very targeted measure as it is most likely to affect heavy drinkers who typically purchase 'cheaper' alcohol products. Although many of us are able to enjoy alcohol responsibly, we are all affected by the small proportion of those that do not or cannot; for example, by the level of disorder you see in our town centres on Friday and Saturday nights, or the associated costs to the NHS and other public services, as well as those who may be quietly drinking themselves into health harms at home. It is a national problem which we all need to face up to.'

In addition to the policy recommendations, NICE has outlined ways that public institutions, such as the NHS, local authorities, police, magistrates, schools and social care services, can improve how they identify adults and adolescents that may be drinking dangerous amounts of alcohol so that they can be put in touch with the relevant support if needed. Among these, NICE advises that:

- Local authorities should consider restricting applications for new alcohol-selling licenses in areas that are 'saturated', if the evidence shows that additional premises could adversely affect public safety, alcohol-related harm, crime and disorder.
- Sanctions, such as fixed penalties and closure notices, should be fully applied to all businesses that sell alcohol to those that are underage or intoxicated, as well as to those that import alcohol illegally. 'Mystery shoppers' have been proven to be an effective way of determining if businesses are doing the best they can.
- Healthcare professionals should use a certified questionnaire to ask their adult and teenage patients about their alcohol intake; for example, during new patient registrations, when screening for unrelated health conditions, when treating minor injuries, and when advising patients about medication or sexual health.

Professor Eileen Kaner, Chair of the Guidance Development group and a Public Health Researcher said: 'There is clear evidence from around the world that we are drinking much more than most other developed countries and the problem is apparent right across society, not just with our teenagers or with binge-drinking.

'We are constantly surrounded by various images of and opportunities to buy alcohol, from promotional offers in supermarkets to advertisements in the media. This encourages us to drink more than we otherwise would, sometimes without us even realising it.

'The guidance from NICE calls for GPs and other public service workers to ask some simple questions about people's drinking habits as early as possible if they ever suspect that there may be a problem. This can help make people aware of the potential risks they are taking or harm they may be doing at an early stage. In addition, we are encouraging practitioners to give brief advice about practical ways of cutting down on heavy drinking. The earlier the problem is tackled, the more likely people are to change their behaviour.'

Ms Jayne Gosnall, community member of the guidance development group said: 'My drinking problem crept up on me when I was going through a bit of a rough time. Rather than facing up to my problems I chose to drink by myself at home, which only made things worse. I constantly put myself at risk by forgetting to lock my doors at night or by accidentally injuring myself. Looking back I think that there were several "missed opportunities", such as when I was in A&E following an accident, where

someone could have asked me a bit more about my drinking habits and whether I was perhaps drinking too much. This might have made a difference and stopped things from getting as bad as they did.

'I hope this guidance from NICE goes some way to increase the opportunities that there are for people, like me, to think about how much they are drinking early on, before it potentially becomes a huge problem for them, their families, and others around them.'

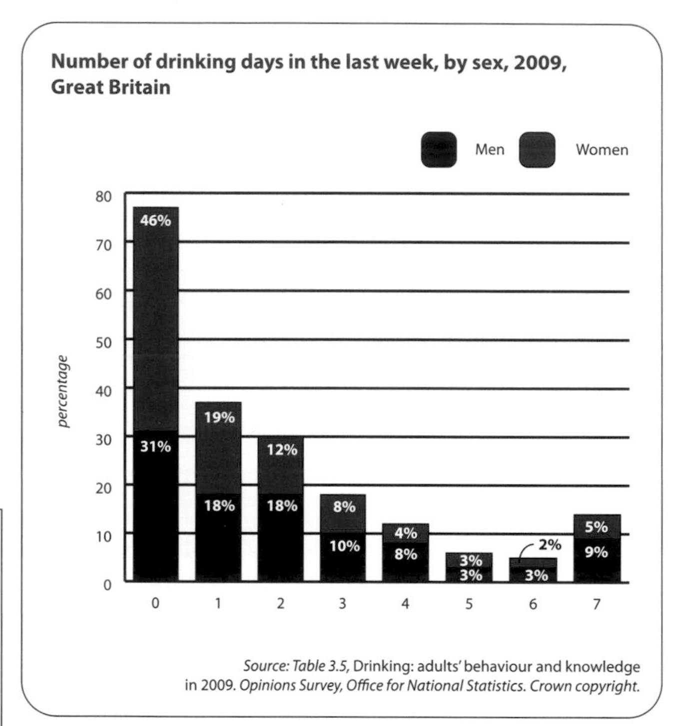

Source: Table 3.5, Drinking: adults' behaviour and knowledge in 2009. *Opinions Survey, Office for National Statistics. Crown copyright.*

Professor Ian Gilmore, President of the Royal College of Physicians and Chair of the Alcohol Health Alliance UK said: 'The nation's increasing addiction to alcohol is placing a huge strain on health services, costing the NHS over £2.7 billion each year. This situation has to change. The NHS's role should not just be about treating the consequences of alcohol-related harm but also about taking early action to prevent alcohol problems, and working in partnership with services in local communities to raise awareness. The Government needs to confront the culture of low prices, non-stop availability and saturation advertising in order to make a difference and cut the amount we are drinking. I support this report from NICE, which presents a timely, evidence-based approach of how to prevent what is fast becoming a public health emergency.'

Professor Alan Maryon-Davis, President of the UK Faculty of Public Health said: 'This is a very practical set of recommendations, based on solid evidence. We need to see this turned into action as soon as possible. The Government should fast track a minimum price to stop the sale of ultra cheap alcohol. And we would like to see licensing authorities restrict the hours of opening for pubs and clubs in city centre trouble spots.'

Professor Steve Field, Chairman of the Royal College of General Practitioners, said: 'The RCGP takes the issue of alcohol abuse very seriously and we have been calling for action for some time. Our manifesto for the recent General Election included demands for minimum pricing and more visible warning labels on all alcoholic drinks so we are pleased to see minimum pricing as a key recommendation in the new NICE guidance.

'Because we work as GPs in the community, we see the dreadful physical and emotional toll of alcohol misuse and so we welcome this guidance. It proposes a series of sensible interventions – such as the alcohol screening questionnaire – that could help prevent serious harm and long-term health problems.

'The clear signposting to other health and support services will also help GPs across the country decide on the right course of action for patients, appropriate to their age and the extent of their problem.'

Key facts about alcohol

- ⇨ 24% of adults drink a hazardous or harmful amount of alcohol.
- ⇨ Between 1980 and 2008, alcohol became 75% more affordable as income rose at a faster rate than the price of alcohol.
- ⇨ 70% of the population who drink responsibly are paying for those who do not as alcohol sold below cost value can only be financed by raising the cost of non-alcohol products.
- ⇨ In 2005, alcohol consumption caused 14,982 deaths.
- ⇨ In 2006/07, alcohol was associated with half a million recorded crimes in England.
- ⇨ Alcohol may be linked to 1.2 million incidences of violence a year.
- ⇨ Up to 17 million working days are lost annually through absenteeism caused by excessive drinking. This costs around £1.7 billion each year, excluding poor performance due to excessive alcohol consumption (such as being hungover).
- ⇨ In 2007/08 there were 863,300 alcohol-related hospital admissions.
- ⇨ Up to 35% of all emergency attendances and ambulance costs are alcohol-related.

2 June 2010

⇨ National Institute for Health and Clinical Excellence (2010). *NICE recommends action to reduce alcohol-related harm.* London: NICE. Available at www.nice.org.uk. Reproduced with permission.

Alcohol advertising

£194 million was spent on alcohol advertising in 2006, nearly half of all food advertising during that year.

Research has shown that advertising can have an effect on the amount a nation drinks and that advertising restrictions can help reduce the appeal of alcohol to young people. This is especially important in the light of the increasing appeal which alcohol advertising has to young people as a 2007 Ofcom study reported. Longitudinal studies indicate that there is a connection between advertising, promotion and the onset and level of alcohol consumption of drinkers.

There are a variety of bodies and organisations that regulate the advertising and promotion of alcohol sales. They include:

- The Advertising Standards Authority (ASA) – regulates all advertising in the UK, with regulations and rulings which are legally binding. They oversee television, poster and radio advertising.
- The Portman Group – regulates the promotion and packaging of alcohol products. Complaints about poor practice and irresponsible products are referred to an independent panel which can order retailers to withdraw a product if it is deemed in breach of the Portman Group code.
- The Cinema Advertising Group – oversee the regulation of cinema advert placement – although the content of adverts still falls under the ASA.

Alcohol Concern has called for a watershed ban on alcohol advertising to limit the exposure of young people to alcohol adverts

There are significant problems with the current regulatory regimes. Each aspect of the regime is industry funded which leads to questions about its impartiality and independence. Further, the means regulators use to assess whether adverts are being shown to under-age audiences are ineffective. Equations to calculate the proportion of young people that may be watching a programme or film are used to determine whether an alcohol advert can be shown. However, these are hard to understand, often are not enforced and there are insufficient consequences if a code is breached.

The financial penalty of having an advert or product withdrawn is a significant deterrent but consumers are ill-informed about who to complain to and even whether a breach of one of the many codes has taken place.

Alcohol Concern has called for a watershed ban on alcohol advertising to limit the exposure of young people to alcohol adverts.

For similar reasons we have called for a ban on all alcohol adverts in cinemas except those with an 18 certificate (there are currently no restrictions about age limits on films and alcohol advertising).

Restricting the way alcohol is promoted and glamorised is an important component in reducing the harm caused by alcohol misuse. The current system of poorly enforced and formed self-regulatory frameworks is a clear case of industry profit being placed ahead of public health.

- The above information is reprinted with kind permission from Alcohol Concern. Visit www.alcoholconcern.org.uk for more information.

DRINK Fizzwells
...FOR THE TOTAL SOPHISTICATE!
THEY DON'T LOOK MUCH LIKE THE PEOPLE I'VE SEEN STAGGERING OUT OF THE LOCAL HERE!

ALCOHOL CONCERN

KEY FACTS

⇨ Overall, the proportion of young people who do not drink is increasing. However, among those who do drink, there seems to have been an increase in alcohol consumption. (page 1)

⇨ 14- to 15-year-olds prefer to drink outside the family environment and are more secretive, hiding their behaviour from their parents. This age group tends to drink to get drunk, with the aim of testing their limits and having fun. (page 2)

⇨ In 2008, 42% of pupils had obtained alcohol in the last four weeks, most commonly by being given it by friends (24%) or parents (22%) or asking someone else to buy it (18%). (page 3)

⇨ How drunk you get depends on how much pure alcohol your drink contains. One way to calculate the amount of alcohol in a drink is by counting units. One unit is 10 ml of pure alcohol – the amount of alcohol the average adult can process within the space of an hour. (page 4)

⇨ After one or two drinks you may start feeling more sociable, but drink too much and basic human functions such as walking and talking become much harder. (page 6)

⇨ Alcohol-related reasons and excuses are responsible for roughly 14 million lost working days in the UK every year. (page 7)

⇨ For men, the recommended limit of alcohol is 21 units a week (3-4 units a day), and for women the recommended limit is 14 units (2-3 units a day). A unit of alcohol is equal to about half a pint of normal strength lager, a small glass of wine, or a pub measure (25 ml) of spirits. (page 8)

⇨ The number of alcohol-related deaths in the United Kingdom has consistently increased since the early 1990s, rising from the lowest figure of 4,023 (6.7 per 100,000) in 1992 to the highest of 9,031 (13.6 per 100,000) in 2008. (page 13)

⇨ It is estimated that three in ten divorces, four in ten cases of domestic violence, and two in ten cases of child abuse are alcohol-related. (page 15)

⇨ More than nine million people in England drink more than the recommended daily amount. More than 9,000 people in the UK die from alcohol-related causes each year. About 20% of these deaths are from cancer, 15% from cardiovascular illnesses, such as heart disease and stroke, and 13% are from liver disease. (page 17)

⇨ During 2007-08, over 30,000 people in England were admitted to hospital with alcohol poisoning – 500 a week – mostly as a result of binge drinking. (page 22)

⇨ Just under one-fifth of children aged 11 to 15 years old have drunk alcohol in the last week, or 558,000 children. (page 25)

⇨ 3.6 million people in Britain have had their childhoods scarred by the drinking of one or both of their parents. (page 27)

⇨ Out of 1,000 adults, 47 are likely to be dependent on alcohol – double the amount of people who are dependent on illegal drugs. (page 28)

⇨ About a fifth of alcohol-related crime is committed in or around licensed premises and there is a link between the density of licensed premises and crime. (page 29)

⇨ Young people are not the worst offenders for excessive drinking, with those aged 35-44 the most likely to drink too much. (page 30)

⇨ More than 810,000 people were admitted to hospital in 2007/8 with alcohol-related illnesses – that's around 1.5 per minute. (page 35)

⇨ £194 million was spent on alcohol advertising in 2006, nearly half of all food advertising during that year. (page 39)

GLOSSARY

Alcohol

The type of alcohol found in drinks, ethanol, is an organic compound. The ethanol in alcoholic beverages such as wine and beer is produced through the fermentation of plants containing carbohydrates. Ethanol can cause intoxication if drunk excessively.

Alcohol By Volume (ABV)

ABV is a measure of how much pure alcohol is present in a drink. It is represented as a percentage of the total volume of the drink. For example, a one-litre bottle of an alcoholic beverage will provide an ABV value on its label. This informs the buyer what percentage of that one litre consists of pure alcohol.

Alcohol dependency/alcoholism

Alcohol is a drug and it is addictive. If someone becomes dependent on drink to the extent that they feel they need it just to get through the day, they may be referred to as an alcoholic. In addition to the various health problems related to alcoholism, an alcoholic's relationships and career may also suffer due to their addiction. They can suffer withdrawal symptoms if they don't drink alcohol regularly and may need professional help from an organisation such as Alcoholics Anonymous to deal with their dependency.

Binge drinking

When an individual consumes large quantities of alcohol in one session, usually with the intention of becoming drunk, this is popularly referred to as 'binge drinking'. It is widely accepted that drinking four or more drinks in a short space of time constitutes 'bingeing', and this can have severe negative effects on people's health.

Depressant

A drug that temporarily causes a decrease or 'slowing down' of the body's mental and/or physical functions.

Drink spiking

When someone adds alcohol or drugs to another person's drink without their knowledge or consent, it is said that their drink has been 'spiked'. Drink spiking is sometimes, but not always, done in order to facilitate another crime such as rape or assault. Prevention strategies include using a 'stopper' in the tops of bottles to prevent anything being added to the drink, and never leaving a drink alone.

Hangover

A hangover describes the effects of alcohol the day after intoxication. Alcohol is a depressant, causes the body to dehydrate and also irritates the stomach, so hangovers usually involve a severe headache, nausea, diarrhoea, a depressive mood and tiredness. There are many myths about how to cure a hangover but the only real solution is to drink plenty of water and wait for it to pass – or of course to drink less alcohol in the first place!

Intoxication

The state of being drunk, caused by over-consumption of alcohol. Drunkeness can lead to dizziness, sickness, loss of memory, aggression or anti-social behaviour, as well as potentially causing long-term health problems such as cirrhosis of the liver. Due to the loss of inhibitions associated with heavy alcohol use, it can also cause people to indulge in risk-taking behaviour they would not normally consider – for example, having unprotected sex.

Teetotal

A teetotaller is someone who abstains completely from alcohol. If an individual is trying to recover from an alcohol dependency they will usually be teetotal, but people do not drink for many other reasons, including religion, pregnancy, for health reasons or just through personal preference.

Unit

The unit system is a method used to measure the strength of an alcoholic drink. One unit is 10 ml of pure alcohol – the amount of alcohol the average adult can process within the space of one hour. Units can be calculated by multiplying the amount of alcohol in millilitres by the drink's ABV, and dividing by 1,000.

Additional Resources

Other Issues *titles*

If you are interested in researching further some of the issues raised in *Responsible Drinking,* you may like to read the following titles in the ***Issues*** series:

- Vol. 188 *Tobacco and Health* (ISBN 978 1 86168 539 1)
- Vol. 186 *Cannabis Use* (ISBN 978 1 86168 527 8)
- Vol. 177 *Crime in the UK* (ISBN 978 1 86168 501 8)
- Vol. 176 *Health Issues for Young People* (ISBN 978 1 86168 500 1)
- Vol. 163 *Drugs in the UK* (ISBN 978 1 86168 456 1)

For a complete list of available ***Issues*** titles, please visit our website: www.independence.co.uk/shop

Useful organisations

You may find the websites of the following organisations useful for further research:

- **Alcohol Concern:** www.alcoholconcern.org.uk
- **The Automobile Association (AA):** www.theaa.co.uk
- **Drink and Drugs News:** www.drinkanddrugsnews.com
- **Drinkaware:** www.drinkaware.co.uk
- **NHS Choices:** www.nhs.uk
- **NICE:** www.nice.org.uk
- **Patient UK:** www.patient.co.uk
- **Priory Group:** www.priorygroup.com
- **TheSite:** www.thesite.org

ACKNOWLEDGEMENTS

The publisher is grateful for permission to reproduce the following material.

While every care has been taken to trace and acknowledge copyright, the publisher tenders its apology for any accidental infringement or where copyright has proved untraceable. The publisher would be pleased to come to a suitable arrangement in any such case with the rightful owner.

Chapter One: Drinking Trends

Young people and alcohol, © Alcohol Concern, *Your guide to alcohol units and measures,* © Drinkaware, *The effects of increasing blood alcohol concentration in a naive male drinker, Alcohol consumption and the risk of physical harm [graphs],* © Medical Council on Alcohol, *Effects of alcohol on your health,* © Drinkaware, *Alcohol facts and trivia,* © TheSite.org, *Drink spiking,* © Crown copyright is reproduced with the permission of Her Majesty's Stationery Office: 2010 – nhs.uk, *Alcohol-related death rates, 1991-2008 [graph],* © Crown copyright is reproduced with the permission of Her Majesty's Stationery Office, *Being teetotal,* © TheSite.org, *Alcohol intake and warning labels [graphs],* © YouGov, *Alcohol deaths,* © Crown copyright is reproduced with the permission of Her Majesty's Stationery Office, *Alcohol and sensible drinking,* © EMIS 2010 as distributed at www.patient.co.uk/health/Alcohol-and-Sensible-Drinking.htm, used with permission, *Social drinking: the hidden risks,* © Crown copyright is reproduced with the permission of Her Majesty's Stationery Office: 2010 – nhs.uk, *Effects of alcohol,* © Public Health Agency, *Young people and alcohol – what are the risks?* © Crown copyright is reproduced with the permission of Her Majesty's Stationery Office, *The 24-hour drinking laws,* © TheSite.org, *Bars with 24-hour drinking laws face 'law and order' levy,* © Telegraph Media Group Limited 2010.

Chapter Two: Problem Drinking

Binge drinking, © TheFamilyGP.com, *Average weekly alcohol consumption by gender, 2009 [graph],* © Crown copyright is reproduced with the permission of Her Majesty's Stationery Office, *Affluent teenagers drink more, study shows,* © Guardian News and Media Limited 2010, *Thousands of children drinking seven pints a week,* © Telegraph Media Group Limited 2010, *Fewer kids drinking – but don't hold the front page,* © Straight Statistics, *Pain and anger are the hidden burden for children with an alcoholic parent,* © Guardian News and Media Limited 2010, *Crime and disorder,* © Alcohol Concern, *Priory research highlights alcohol problem,* © Priory Group, *Drink driving,* © Automobile Association, *Half of children see parents drunk,* © Press Association Ltd 2010, *Cutting drink-drive limit could save lives, NICE study finds,* © National Institute for Health and Clinical Excellence, *GPs to quiz patients over alcohol intake,* © 2010 Associated Newspapers Ltd, *Knowledge of drinking guidelines does not equal sensible drinking,* © 2010 Elsevier Ltd, *The real cost of alcohol,* © Drink and Drugs News, *NICE recommends action to reduce alcohol related harm,* © National Institute for Health and Clinical Excellence, *Number of drinking days in the last week, by sex [graph],* © Crown copyright is reproduced with the permission of Her Majesty's Stationery Office, *Alcohol advertising,* © Alcohol Concern.

Illustrations

Pages 5, 14, 24, 32: Angelo Madrid; pages 8, 20, 27, 34: Simon Kneebone; pages 11, 22: Bev Aisbett; pages 19, 29, 36, 39: Don Hatcher.

Cover photography

Left: © Dawson R. Toth. Centre: © Andrew Richards. Right: © Maryke de Beer (marykedbeer@gmail.com).

Additional acknowledgements

Research and additional editorial by Carolyn Kirby on behalf of Independence.

And with thanks to the Independence team: Mary Chapman, Sandra Dennis and Jan Sunderland.

Lisa Firth
Cambridge
September, 2010